FOR REFERENCE

Do Not Take From This Room

Date: 8/24/04

PALM BEACH COUNTY
LIBRARY SYSTEM

3650 Summit Boulevard
West Palm Beach, FL 33406

A TO Z
OF
MARINE SCIENTISTS

BARBARA CHARTON

Facts On File, Inc.

A TO Z OF MARINE SCIENTISTS

Notable Scientists

Copyright © 2003 by Barbara Charton

Facts On File, Inc.
132 West 31st Street
New York NY 10001

Library of Congress Cataloging-in-Publication Data

Charton, Barbara
    A to Z of marine scientists / Barbara Charton.
       p. cm.—(Notable scientists)
    Includes bibliographical references and index.
    ISBN 0-8160-4767-7 (hardcover: acid-free paper)
    1. Marine scientists—Biography—Dictionaries. I. Title. II. Series.
    GC9.C48 2003   551.46'0092'2—dc21
    [B]2002152445

Facts On File books are available at special discounts when purchased in bulk quantities for businesses, associations, institutions, or sales promotions. Please call our Special Sales Department in New York at (212) 967-8800 or (800) 322-8755.

You can find Facts On File on the World Wide Web at http://www.factsonfile.com

Text design by Joan M. Toro
Cover design by Cathy Rincon
Chronology by Dale Williams

Printed in the United States of America

VB FOF 10 9 8 7 6 5 4 3 2 1

This book is printed on acid-free paper.

CONTENTS

LIST OF ENTRIES

ACKNOWLEDGMENTS

My thanks go to Marvin (Great Mind), who was always helpful, and to my many and varied friends and patient children who endured me.

Heartfelt thanks are due to the Grolier Information Service (and our fearless leader, Sheila MacDonald), where I learned both to do research and to love it.

A final note to the staff of Facts On File and to Ms. Elizabeth Oakes for their dedication to the publication of quality reference material.

INTRODUCTION

Marine science is not a single discipline. It is a body of knowledge about aquatic biomes that has been assembled by researchers who come from many, often interrelated disciplines. Some marine scientists are biologists, ecologists, paleontologist, zoologists, and in the past, natural scientists. Others were trained as chemists, geologists, meteorologists, and physicists. Work by the navigators and cartographers added to and will continue to augment the efforts of the others.

The connections are numerous and sometimes startling: bits of information gathered in one field may lead to significant insights for another researcher. The work of the early enthusiasts, people who were studying science for their personal entertainment, is not to be discounted. The body of knowledge that they bequeathed to their successors was formidable. As new instruments refine measurement, and new perspectives emerge, the same data—or a transmuted form of it—is used to formulate different ideas. Marine science is not a closed discipline. We do not know everything about the ocean or its nearest neighbors, the brackish wetlands, the adjoining coasts, or the ocean of air above them. This is the frontier of marine science.

THE SCIENTISTS

A to Z of Marine Scientists brings together an array of marine scientists, providing basic biographical details of their lives. The focus, however, is on their work, with their scientific achievements situated in their proper social context and presented in everyday language that makes even the most complex concepts accessible.

Because the field of marine science is interdisciplinary, A to Z of Marine Scientists presents scientists from backgrounds as diverse as archaeology, biology, geology, and oceanography. No one discipline dominates.

Some are well-known scientific greats; others are contemporary scientists, whose work is just verging on greatness. Among these are minority scientists and inventors, who have often been excluded from books such as this.

THE ENTRIES

Entries are arranged alphabetically by surname, with each entry given under the name by which the entrant is most commonly known. The typical entry provides the following information:

Entry Head: Name, birth/death dates, nationality, and field(s) of specialization.

Essay: Essays range in length from 250 to 2,000 words, with most averaging around 1,000 words. Each contains basic biographical information—date and place of birth, family information, educational background, positions held, prizes awarded, etc.—but the greatest attention is given to the entrant's work. Names in small caps within the essays provide easy reference to other people represented in the book.

In addition to the alphabetical list of scientists, readers searching for names of individuals from specific countries can consult the appendix in the back of the book, which indexes entrants by country of birth and/or citizenship. The Chronology lists entrants by their birth and death dates. Finally, the main index lists page references for scientists and scientific terms used in the book.

A

Agassiz, Alexander
(1835–1910)
American
Oceanographer

Alexander Emmanuel Rodolphe Agassiz was one of the people referred to as "the father of oceanography" by other oceanographers. While that is an unofficial title, he has achievements that make this a reality. Agassiz was born in Neuchâtel, Switzerland, on December 17, 1835.

The son of a famous father, LOUIS AGASSIZ, Alexander and his two siblings moved to the United States almost two years after their mother, Cécile Braun, died, and his father had remarried. They arrived in 1849, and the elder Agassiz prepared his son for entrance to Harvard College. He graduated in 1855 and then entered the Lawrence Scientific School, the scientific division of Harvard, since science courses were not then part of the regular Harvard College curriculum. At Lawrence, he studied engineering and chemistry, and graduated in 1857.

His father envisioned Alexander as his heir apparent in Cambridge, Massachusetts, but the younger Agassiz joined the United States Coast Survey as an assistant, and this post took him to California in 1859. The Survey's project of mapping the western coast of North and Central America took him down the Pacific coast to Aca-pulco and eventually to Panama. While traveling down the Pacific coast of Central America, he gathered marine specimens for his father's museum collection. On his return to San Francisco, he worked as a mine inspector. After this year in the west, he returned to Cambridge once again.

The year 1860 marked several beginnings for Alexander. It was then that he married Anna Russell. Since he felt undereducated in some fields, Agassiz enrolled once again at Lawrence to study zoology and geology. The result of his many areas of expertise made him the obvious candidate for a natural history post at a major museum. It meant that he was conversant with the work of all the departments. He was appointed assistant curator in the Harvard (Cambridge) Museum of Natural History in 1865. By 1870, during his father's absence on another collecting voyage, the younger Agassiz was appointed curator and remained a presence in the museum until he died.

Throughout the 1860s, he worked with other family members in mining, both in Pennsylvania and Michigan. This endeavor earned him a private fortune and freed him from the need to earn other money. He devoted both his time and his funds to science, donating large sums and his collections to Harvard College and the natural history museum. After Anna Agassiz died in 1873, Alexander spent increasing time in scientific pursuits. His definitive revision of *Echini* (Echinoderms—starfish and their numerous relatives) was published in fascicles, or sections, between 1872 and 1874. In the work, he used the embryological and paleontological techniques developed by his father. This work was intended to be a support of his father's position as a notable antievolutionary biologist, and another refutation of CHARLES DARWIN. The son felt that he had to uphold his father's view of man's special place in creation. In spite of his declared position, the work pointed to a Darwinian explanation of echinoderm species diversification and spread.

Agassiz had maintained his connection to the U.S. Coast Survey. The *Blake*, a steamer, was placed at his disposal. He used the ship for deep-sea dredging during the winters of 1876–81. The *Albatross*, a schooner that belonged to the U.S. Fish Commission, was used for several other voyages, notably one where he was accompanied by his stepmother, to South America. The scientific part of that trip was coastal surveying and specimen collection. The Fish Commission had charged the expedition with mapping areas of fish schooling. This was of immediate commercial importance.

Some of Agassiz's expeditions were at the behest of the Coast Survey, several to the Caribbean. His mission there was to chart the coasts and the path of the Gulf Stream, and to measure the velocity and paths of the currents in the Caribbean. This was of great interest to the shipping industry and the growing telephone communications network that was bridging continents. Agassiz also returned to the Pacific, where one voyage took him to the Great Barrier Reef off the Australian coast. Again, he was surveying the oceanic biomes, estimating both the number of species and the populations of each species.

In 1899 he went deep-sea dredging in the Pacific, off the Marquesas. This voyage was again at the direction of an arm of a governmental agency. The charge there was a scientific comparison of Pacific species. Fossil sharks' teeth were retrieved by the dredges at great depth. Since

sharks are relatively shallow-water animals, Agassiz reasoned that these remains could only be present at the depths where they were found if the ocean floor had subsided. His explanation was bringing him closer and closer to Darwinism and the notion of an ever-changing Earth. These conclusions were supported by later geological evidence.

As one of the foreign scientists who were called upon to help classify the enormous number of specimens brought back to England by the HMS *Challenger* in 1875, Agassiz worked on the Echinoderms. The project occupied him for several years, and the connection to researchers in Great Britain continued. Agassiz was aboard ship, traveling from Great Britain to the United States when he died on March 27, 1910.

The links between various disciplines that all touch on marine science were made clear by Alexander Agassiz. A student of many scientific fields, he could see where they came together to create oceanography as a discipline that draws from the many different sciences that comprise it. As a research team leader, he inspired and coordinated the efforts of many, and as a benefactor of science and an enthusiast who helped convince the public that such work was necessary, he was incomparable.

⊠ **Agassiz, Louis**
(1807–1873)
Swiss/American
Natural Historian, Geologist

Louis Agassiz made major contributions to both of his scientific fields on two continents. He was born Jean-Louis-Rodolphe on May 28, 1807 in the alpine town, Môtiers-en-Vuly, in the Swiss canton of Fribourg. His parents, Rodolphe and Rose Mayer Agassiz, expected that Jean Louis would follow his father in the ministry, but the young man preferred a medical career. Students then did not expect to spend their university time in one school; following this pattern, Agassiz was educated at the universities in Zurich, Heidelberg, and Munich. His medical degree was granted in Munich in 1830. While still a student, he became increasingly interested in natural sciences, the growing vogue among university students of his day.

Excited by a report of an expedition to Brazil that brought back a large number of fish unknown to Europeans, he wrote a monograph on the subject. That paper impressed GEORGES CUVIER, the internationally recognized expert on anatomy and paleontology, who invited the young doctor to join him in Paris and follow his interest in natural history. This was a much sought-after honor. While in Cuvier's research group, he produced the definitive catalog of the fish specimens brought back from Brazil by C. F. P. von Martius and J. B. von Spix, in which he attempted to classify the fish. The elaborate scheme that Agassiz devised was based on differences in scales. Although it is not used now, this was a major undertaking for its day.

During his stay in Paris, Agassiz met FRIEDRICH VON HUMBOLDT, who recommended him for the post of professor of natural history in the newly opened university in Neuchâtel, Switzerland. With Cuvier's recommendation, he began his career there in 1832 and married Cécile Braun, the sister of a Fribourg classmate, in the same year.

During Agassiz's tenure at Neuchâtel, he published two major works. In an extensive study of glaciers, *Étude sur les glaciers* (Study of glaciers), published in 1840, he discussed glacier characteristics, the rocks, minerals, and fossils found in glaciers, and their movement down mountains. The term *Ice Ages* to describe the periods in Earth's history when glaciers covered extensive areas is attributed to Agassiz. When describing Earth's history, he used the conventional explanation for fossils; he was a catastrophist in Cuvier's tradition. In catastrophist theory, all life forms were individual creations, and their catastrophic destructions were followed

Louis Agassiz *(E. F. Smith Collection, Rare Book & Manuscript Library, University of Pennsylvania)*

by new creations. Noah's flood was one such catastrophic event.

The other work of Agassiz's Neuchâtel tenure was his masterpiece, *Récherches sur les poissons fossils* (Research on fossil fish). This was a major undertaking consisting of five volumes and an atlas, issued in parts between 1833 and 1844. The illustrations were done by his wife, Cécile Braun, an accomplished artist.

By the time he was 40, Louis Agassiz was an international celebrity—he was "Monsieur Science" to the general intellectual public. A suggestion by CHARLES LYELL, the British geologist, in 1846 took Agassiz to the United States the next year on a natural history lecture tour. The tour was a success; he proved to be a popular and engaging speaker who made the subject come alive. During the next, momentous year, his wife died of tuberculosis, the tour ended, and Harvard offered him a position. Accepting the post, he began teaching at the Lawrence Scientific School—the science division of Harvard. The American phase of his life had begun. Agassiz quickly became a part of the intellectual life of Cambridge, Massachusetts. Henry Wadsworth Longfellow and Oliver Wendell Holmes were his colleagues at Harvard and fellow members of the Fortnightly Club. Holmes wrote *The Chambered Nautilus* for one of Agassiz's birthday celebrations. Elizabeth Cabot Cary married Agassiz in 1849. She was a part of his naturalist circle and the catalyst that piqued his interest in women's education. Louis taught in the newly emerging schools for women, and Mrs. Agassiz eventually became the founder and first president of Radcliffe College. The three children of his first marriage moved to Cambridge in 1849.

Agassiz continued research, specimen collecting, and popular lectures on natural history. He organized and sailed on two expeditions exploring the American coasts: the Atlantic coast in 1871 and the Pacific coast the next year. A vast 10-volume work, *Contributions to the Natural History of the United States*, was planned. Four volumes appeared between 1857 and 1862, but this great survey was never finished. The published material does contain Agassiz's *Essay on Classification*; this is a lengthy refutation of the theory of evolution expounded by CHARLES DARWIN. In this work, Agassiz maintains his traditionalist's position on evolution: mankind is unique, God-created, and arranged in different species. Traumatic changes in Earth's history created diversification, according to Agassiz. This argument, coming from a recognized scientific celebrity, was used as a rationalization for slavery. Agassiz spent much of his time after 1860 arguing against the Darwinian evidence for natural selection.

Using his considerable influence, Agassiz raised huge sums to fund scientific research in the United States. One of the recipients was the Museum of Comparative Zoology at Harvard, which is now named for him. He helped establish the National Academy of Sciences in 1863, in conjunction with ALEXANDER BACHE, and continued to support the work of the U.S. Coast Survey and of the Smithsonian Institution, the national natural history research organization. His last collecting trip took him to South America and through the Straits of Magellan. He died of a cerebral hemorrhage on December 14, 1873, shortly after the voyage.

⊠ **Albert I**
(Albert-Honoré Grimaldi)
(1848–1922)
Monegasque
Oceanographer, Meteorologist,
Museum Founder

Prince Albert I of Monaco was born in Paris on November 13, 1848. His parents were Charles III, prince of Monaco, and Antoinette Ghislane. He grew up in France and joined the Spanish navy as a young ensign. During the Franco-Prussian War in 1870, he was a lieutenant commander in the French navy. His involvement with the sea seemed inevitable to him because Monaco is perched on the Mediterranean.

The first of his scientific ships, bought in 1873, was a schooner, the *Pleiad*, renamed *Hirondelle* (Swallow). This vessel sailed on four voyages. There were eventually three more ships: another *Hirondelle*, and the *Princesse Alice I* and *II*. The scientific equipment and the capabilities of the scientific crew increased with each change of vessel. In all, Albert spent 40 summers at sea.

Together with his group of scientific colleagues, Albert studied ocean currents in the Atlantic, specifically the Gulf Stream, and charted drifts in the North Atlantic and the Arctic. They systematically sampled water, studying temperature at varying depths and locations, as well as chemistry. The prince's specialty was meteorology, important to anyone who used the ocean. Study of

Albert I, prince of Monaco *(Hulton Archive/ Getty Images)*

Atlantic storm movements was always a concern for European sailors and fishermen. To increase available information on storm movements, Albert established four weather stations in the Azores as an early-warning and tracking network.

Dredging and the identification of specimens retrieved was a part of the biological research of each voyage. The prince directed deepwater dredging to depths of over 6,000 m (20,000 ft) and plankton sampling from the surface to depths of 5,000 m (16,500 ft). Albert is also credited with supporting the research leading to the isola-

tion and identification of the neurotoxin in the stinging cells of *Physalia*, the Portuguese man-of-war. This substance induces deep anesthesia or death, depending on the size and bodyweight of the victim.

"Monsieur le Prince," as his colleagues called him, supported many organizations. He sponsored the Institut océanographique, Institut de paleontologie humaine, and the public Musée anthropologique in Paris and the Musée océanographique in Monaco. This museum houses his private collection and a wonderful aquarium.

The findings from the scientific voyages were published in a journal founded by Albert. It continues after several name changes. The original was *Bulletin du Musée Océanographique*. This became *Bulletin de l'Institut Océanographique* and then *Annales de l'Institut Océanographique*. He also published in *Comptes rendus des Séances de la Société de Biologie* and *Bulletin de la Société de Géographie*.

Albert urged intelligent use of the ocean's resources. He was an early spokesman for limits on the capture of whales and other animals hunted commercially, warning any group that would listen, and publish his findings, about declining populations of certain species. He also spoke tirelessly for clean use of the seas and warned about the dangers of pollution and its effect on the oceanic biome.

In addition to his memberships in various scientific societies and the *Academie des sciences*, Albert was a member of the *Academie d'Agriculture*. One of his estates was a farm near Paris. There, he implemented modern farming methods and experiments. He died there on June 26, 1922.

Andrusov, Nikolai Ivanovich
(1861–1924)
Russian
Geologist

Andrusov was born in Odessa, Russia, on December 19, 1861. His parents were Ivan Adreevich and Elena Filipovna Belaya. His father was a navigator and died in a shipwreck in 1870 when the boy was young. Andrusov attended the local schools and became interested in the local geology while a student in the Kerch Gymnasium. He maintained that interest as a university student at Novorossiysk University and was so obviously a star pupil that the Society of Naturalists of the university sent him on a summer collecting trip to the nearby Kerch peninsula.

While an undergraduate, he had signed a political protest letter, and in spite of his qualifications in geology, he was not appointed to the faculty. He went abroad instead, studied in Vienna and Munich, and went on geological surveys in the Tyrol, Croatia, and Italy. He returned to Russia and to a position in St. Petersburg at the university in 1886. The Petersburg Society of Naturalists in the following year sent him to the Trans-Caspian region to explore the area and the next summer again to the Kerch peninsula. There he was concentrating on the hydrogeology of the region.

In 1889, he married Nadezhda Genrikhovna, the daughter of Heinrich Schliemann, the discoverer of the ruins of ancient Troy. He advanced academically, receiving his M.A. at Petersburg University. His thesis topic was based on his researches of the Kerch limestones. This led to an appointment in geology at Novorossiysk University. The Russian Geographical Society equipped the ship *Chernomorets* for him. The object was an exploratory cruise to study the depths of the Black Sea. While there, he was involved in the establishment of institutions of higher learning for women. Andrusov then taught geology for women in the school he helped found. In 1893, the family moved back to St. Petersburg; Andrusov was appointed lecturer in paleontology. He spent the next three summers cruising the Dalmatian coast, the Sea of Marmara and the waters off Baku, measuring depths and sampling sediments. Another move was necessary when he became professor of paleontology and geology

in Kiev. During World War I, he directed a commission responsible for studying Russia's natural resources. After the war, he suffered a stroke, and his wife insisted that they move to Paris, where he recuperated. Because his health had improved, he then lectured on geology at the Sorbonne as a visiting professor. He retired again and moved to Prague, where he died on April 27, 1924.

Andrusov was a major figure in Russian oceanography. His reputation is based on significant work studying fossil reefs in the Kerch peninsula, the organisms and limestone they deposited, and the tectonic structures and submarine terraces of this geologically active region. His work on the Neocene period in southern Russia was an extensive study of the fossil fauna and biogenesis of that fauna and its environment. Andrusov recognized long before others did that the environment organisms live in is of extreme importance to their continued existence.

⊠ Anning, Mary
(1799–1847)
English
Naturalist, Fossil Hunter

Mary Anning lived all her life in Lyme Regis on England's southern coast. Anning was born on May 21, 1799; her parents, Richard and Mary, had 10 children whose births were recorded in the local church. The children spent several years in local schools, where they were instructed in basic subjects, but all the children were introduced to their father's hobby of fossil hunting. The entire family joined in this dirty and often dangerous avocation, but Mary was the most adept. Fossil hunting was a difficult and often dangerous pursuit. These amateurs dug away at the eroding cliff-sides of Lyme Bay using hand tools and trying not to create earthslides.

Richard Anning was a cabinetmaker who was barely supporting his large family. When he died in 1810 the family was penniless. They

Mary Anning was born on the site of the Lyme Regis Philpot Museum. The fossils she collected, including several creatures unknown at the time, provided the basic material for her exploration of the science of geology. *(The Natural History Museum, London)*

earned what they could by the sale of fossils to collectors. This business brought Mary to the attention of Thomas Birch, a gentleman and a wealthy fossil collector, who sold part of his own collection to support the family.

The first specimen of *Ichthyosaurus*, the first almost-complete fossil of *Plesiosaurus*, and the first British *Pterodactylus macronyx* (a flying reptile) are among the finds credited to Mary Anning. Some of these are now in the Natural History Museum in London. Anning was recognized as an authority on anatomy and paleontology by her contemporaries. Many considered her a scientific outsider and therefore suspect: she was not male, educated, or aristocratic, and her specimens were sold, rather than donated to mu-

seums. Some scientists refused to believe that a female and a scientific nobody could produce significant work; GEORGES CUVIER, the French anatomist and famous palentologist, was one such scientific notable. He labeled her plesiosaur a fraud but acknowledged that she was a fine anatomist.

With increasing recognition, she was granted an annuity by the British Association for the Advancement of Science (1838) and a sum by the Geological Society of London. When she died of breast cancer on March 9, 1847, her obituary appeared in the *Quarterly Journal of the Geological Society*. This organization did not admit women to its membership until 1904.

Aristotle
(384–322 B.C.E.)
Greek
Philosopher, Natural Scientist

The famous world traveler, scientist, and teacher of Alexander the Great came from Khalkís. His family was one of physicians. Although his father died when the future philosopher was a boy of about 10, as a child, he did have exposure to the pharmaceuticals of the era. These were all natural products derived from plants, and physicians were expected to make their own remedies. Aristotle is first noted in Athens when he was about 17 years old. Starting as a student in Plato's Academy, he remained there for 20 years. After Plato's death and a political change, Aristotle left Athens in 342 B.C.E. for an extended stay in Ionia and Lesbos, where he studied biology. Returning to mainland Greece, he taught the young prince, Alexander, in Macedon and went back to Athens with him. After Alexander's death in 323 B.C.E. and the resurgence of Alexander's enemies, Aristotle was charged with impiety. He chose voluntary exile to Khalkís and died shortly thereafter.

Aristotle's work on what might be called physical oceanography, the *Meteorologica*, is in

A sculptor's vision of Aristotle *(Synnberg Photo-gravure Co., Chicago, Philosophical Portrait Series, Open Court Publishing Co., Chicago, courtesy AIP Emilio Segrè Visual Archives)*

part a compilation of the works of others. He did state firmly that the water in the sea was evaporated by the Sun and eventually returned to Earth as rain, which created streams and rivers. His explanation of the saltiness of the sea was not one that we now hold: he believed the salt to be "an exhalation of earth." This was based on his observation of the saltiness of some south winds. This is common; windblown salt still comes from the sea.

The importance of Aristotle's work in physical oceanography is that he used observation, and based his theories of rainfall and tides on a belief that these phenomena were natural events and could be logically explained. While the movement downward of water in rivers that eventually terminated at the sea was plain, the idea of tides was troublesome. Aristotle had an advantage over other Greek philosopher-scientists; he traveled widely. Most of their work was based on the tides in the Mediterranean, and these tides are not very marked; there are relatively small differences between high and low water. The idea that there is one cause for all tidal movement, a natural cause at that, is a sophisticated one. A 17th-century writer attributes Aristotle's drowning death to suicide. He was supposedly frustrated with his futile attempts to explain tidal variation.

Aristotle's writing on biology came in the latter part of his life. Here, too, he compiled information from other sources but applied logic to travelers' tales of wondrous animals. He also did his own careful observations; thus, careful examination of leg joints could then accurately establish just how a tetrapod—notably the elephant—walked. Until his time, it was assumed that the elephant had no leg joints, and the animal slept standing against trees.

While on Lesbos, he obviously did dissections, particularly of marine creatures, notably squid, octopuses, cuttlefish, sea urchins, scallops, and sponges. Careful observation and description of the mouthparts of the sea urchin have come down to us, as well as the explanation of the ability of the cuttlefish to anchor itself on rocks in storms. Aristotle observed the species-specific egg-tending behavior of male catfish, and tested sense perception in scallops and sponges. His study of eggs, urchin's eggs, bird eggs, fish eggs, what would now be called embryology, was part of his attempt to classify living organisms. Embryological study was part of Aristotle's approach to the concept of what makes a living creature alive. He looked at the increasing complexity and equated that with a "scale of Nature," or increasing "soul." His human physiology, since he did no human dissection, was poor.

Much of Aristotle's biological observations were either ridiculed or ignored in later time. It is only in the last century that his carefully detailed studies are coming into full recognition. His influence on the thinking behind European and Arabic science is incalculable. He was an early expounder of the "hands-on" approach to biology: this was a science that required experimentation and dissection. Many of his observations were not accepted until the 19th century when his carefully notated work was corroborated.

His attempts at classification were important because they provided a beginning. The Muslim scientific community tried to continue them, but the most important consequence of Aristotle's classification scheme is that it was the prototype for the 18th-century works that finally succeeded.

⊠ Audouin, Jean-Victor
(1797–1841)
French
Biologist, Physiologist

Audouin was born in Paris on April 27, 1797, to Victor-Joseph and Jeanne-Marie Enén. His father was in the diplomatic service, and the family moved frequently. Audouin was educated in Rheims, in Paris, and then in Lucca, Italy, before returning to Paris and the prestigious secondary school, Lycée-Louis-le-Grand. He began in law, but soon switched to medicine, pharmacy, and natural sciences. His first published works in 1816 on insects brought him to the attention of GEORGES CUVIER, who was undoubtedly the most famous French scientist of the early 19th century. This was an enormous advantage to Audouin, since it brought him into the forefront of natural science research.

Audouin was one of the founders of the Société de la histoire naturelle de Paris (Natural History Society of Paris) in 1822, and of the journal, *Annales des sciences naturelles*, in 1824. He was one of the editors of this periodical. Recognition for these contributions came when he was appointed as assistant to Jean-Baptiste de LAMARCK, zoologist at the Museum of Natural History in Paris.

Two years later, Audouin and HENRI MILNE-EDWARDS began a collaborative encyclopedic work on the anatomy and physiology of the marine invertebrates of the northwestern coast of France. As a result of these studies, he published a major work on the crustacea, *Récherches anatomiques et physiologiques sur la circulation dans les crustacés* (Anatomical and physiological research on the circulation in crustaceans). In 1827, Jean-Victor married his draftsperson and collaborator, Mathilde Brongniart. Her brother was one of Jean-Victor's coeditors of *Annales*.

The year 1832 encompassed several events. With Milne-Edwards, Audouin published the first volume of *Récherches pour servir à l'histoire naturelle du littoral de la France* (Research to be used in the natural history of the French coast). Carolus Linnaeus's binomic classification of coastal invertebrates was introduced in this work. This was a major taxonomic advance since most invertebrates up to this time had been lumped together as "worms." Also in 1832, Audouin was one of the founders of the Societé entomologique de France (Entomological Society of France). By the next year he was professor of zoology and studying arthropods in general.

Audouin spent the remainder of his life in the study of insects and applied entomology. He died suddenly on November 9, 1841, and his last works were finished and published for him by Milne-Edwards.

⊗ **Bache, Alexander Dallas**
(1806–1867)
American
Physicist

This well-known physical scientist was born into a family of distinguished scientists. The eldest child of Sophia (Dallas) and Richard Bache was born on July 19, 1806, a great-grandson of Benjamin Franklin. The family was a distinguished one in Philadelphia. Young Alexander was educated in Philadelphia, and received his formal education at West Point. After graduation in 1825 he served in the Corps of Engineers. Upon his return to Philadelphia, he was associated with the Franklin Institute there, and then taught mathematics and physics at the University of Pennsylvania and Girard College. Bache was a member of the group that organized Girard, and he was its first president. Alexander Bache married Nancy Clarke Fowler in 1828.

From 1843 to 1861, Bache was superintendent of the United States Coast Survey. He was active in the charting of the Texas coast in cooperation with the Texas Navy (Texas at this point was an independent country) and most of the gulf coast to Florida. During the Civil War, Bache was an adviser to the United States Navy. With Joseph Henry, of the Smithsonian Institution, Bache created early links between research insti-

Alexander Bache was a charter member of the National Academy of Sciences and its first president. *(E. F. Smith Collection, Rare Book & Manuscript Library, University of Pennsylvania)*

tutions and government. They saw the need for a national science organization, and together with LOUIS AGASSIZ and others, formed the National Academy of Sciences in 1863, with Bache as the first president of the group. After suffering an incapacitating stroke in 1864, Henry oversaw the academy and, after Bache's death on February 17, 1867, the organization was financed by Bache's bequest.

During his tenure at the Coast Survey, in less than 20 years Bache made it the largest employer of physical scientists in the nation. The Survey helped develop accurate scientific maps of the entire United States coastline, charted the Gulf Stream, and also made careful observations of the waves produced by a Japanese earthquake; this led to the department's research in what became the science of seismology. Bache left a legacy of sound work in marine physical geography that was important to his nation.

Bailey, Sir Edward Battersby
(1881–1965)
British
Geologist

Edward Bailey was born in Marsden in eastern England on August 1, 1881, to Louise (Carr) and John Battersby Bailey. He did not follow his father into medicine but went to Cambridge, where he received a degree with highest honors in physics and geology in 1902. A first position with the Geological Survey took him to Scotland. The Geological Survey is a government agency that creates and maintains extremely detailed large-scale maps of the entire landmass of Great Britain. Most of his research was concentrated on the geologic structure and rock forms of Scotland. During World War I, he was in the artillery, was wounded, and lost an eye. He was awarded the Military Cross by Great Britain and the Croix de Guerre and the Legion of Honor by France.

After a brief period as professor of geology at the University of Glasgow, Bailey returned to the Survey as its director. He pursued studies of mountain systems and their shifts: the jumbled mountains of Scotland provided him with a laboratory in which to observe the extensive folds and buckling of the landmass, supporting the theory of continental drift. The extensive work in the 1930s on submarine landslips in Sutherland, Scotland, gave others, notably PHILIP KUENEN, the material with which to develop their theories explaining turbidity currents. These in turn are a major component of the explanation of submarine canyon shape. All of this work supported the theory of plate tectonics.

Bailey wrote for both professionals and amateurs interested in geology. For a professional audience, he wrote on the history of geology and the development of tectonic research. Among his writings aimed at a general public were biographies of CHARLES LYELL and James Hutton, British pioneers in geology. Some of his writings that deal directly with marine science are "The

Edward Bailey *(University Archives, University of Glasgow)*

Desert Shores of Chalk Seas" published in the *Geological Magazine* (1924), "Paleozoic Submarine Landslips near Quebec City" in the *Journal of Geology* (1928), "New Light on Sedimentation and Tectonics" in the *Geological Magazine* (1930), and a well-received paper, "Submarine Faulting in Kimmeridgian Times" in *Transactions of the Royal Society of Edinburgh* (1932).

Bailey was elected a fellow of the Royal Society of London, and he was knighted in 1945, which was also the year of his retirement. His popular writings occupied his working time in retirement, and he died in London on March 19, 1965.

⊠ Ballard, Robert
(1942–   )
American
Marine Geologist, Explorer

Robert Ballard was born in Wichita, Kansas on June 30, 1942, and grew up in California. When he was a child, the sea fascinated him, and this interest continued. It led him to the University of California and a degree in geology and chemistry in 1965, then to a year studying marine geology and oceanography at the University of Hawaii in 1966, another course of study at the University of California in geophysics and geology in 1967, and a Ph.D. in marine geology from the University of Rhode Island in 1974. He then did postgraduate work in Hawaii. Ballard joined the U.S. Navy in 1967. He was assigned to the Deep Submergence Laboratory in Woods Hole, Massachusetts, and remained there until 1997.

The general public became aware of Ballard because of his spectacular, newsworthy deep dives, notably to the sunken RMS *Titanic* in 1985. That action, however, was just one part of a long career using deepwater submersibles. The *Alvin*, a small submarine, was the vessel used in many dives to the Mid-Atlantic Ridge in 1973–74. Another site he explored, near the Galápagos, was the first

Robert Ballard, president of the Institute for Exploration in Mystic, Connecticut, announces the discovery of eight ships, deep in the Mediterranean, on July 30, 1997, at National Geographic. The discovery is the largest concentration of ancient shipwrecks ever found in the deep sea. More than 100 artifacts were recovered from the wrecks. *(Photo by Laura Camden REUTERS/Getty Images)*

place where anyone saw the huge tube worms and other fantastic organisms living on the "black smokers," the structures that are deposited by undersea thermal vents. Ballard then found similar creatures in other underwater "hotspots," near Baja California and in the Caribbean.

Ballard's vessel, the *Alvin*, was refitted and modified to expand its range and make it more comfortable. A robot, the JASON, was designed to do the actual exploration of large and possibly unsteady wrecks—the *Titanic* was one where the JASON was used. Underwater sleds have also been designed and used.

In addition to exploration and production of a considerable body of scientific and popular literature, Ballard was the founding chairman of the JASON Foundation for Education. This group brings students almost as close to live exploration as they could be without being in the submersible. The students see the dives via live television as they are progressing. Thousands of students have witnessed, almost firsthand, underwater volcanoes, coral reefs, shipwrecks, and animal habitats in locales as varied as the Sea of Cortez, the Mediterranean, the Florida Keys, the Belize coast, and the Great Lakes.

Because of his work, Ballard has been awarded honorary degrees by a number of universities. He is a medallist and received the Newcomb Cleveland Prize and the Westinghouse Award from the American Association for the Advancement of Science, the Cutty Sark Award from Science Digest, and the William Proctor Award from Sigma Xi. He continues exploration and study emphasizing advanced deepwater technology and techniques, including the used of submersibles, and the mid-ocean ridges.

George Bass (left) examining pottery from the fourth-century wreck of the *Yassiada*. *(Institute of Nautical Archaeology)*

⊠ **Bass, George**
(1932–)
American
Marine Archaeologist, Explorer

George Bass is credited with being the founder of nautical archaeology. Bass was born in Columbia, South Carolina, and began his college education at Johns Hopkins University, where he soon abandoned his original major, English, for archaeology. He received his M.A. in 1955. As a doctoral student at the University of Pennsylvania, he was asked to direct the excavation of a Bronze Age ship found in relatively shallow water off the Turkish coast. He took diving lessons at the YMCA and went to Turkey.

The site was 90 feet down. It had first been found in 1954 by a sponge diver and was later brought to the attention of the University of Pennsylvania by Peter Throckmorton, a photog-rapher-adventurer who maintained an interest in what he considered "his" shipwreck for years. A team of French, American, and Turkish archaeologists assembled to investigate the wreck. They determined it to be a merchant vessel that foundered about 1200 B.C.E. and was obviously not Mycenaean, as all seafaring ships were then assumed to be. It was from the Levantine coast and carried a cargo of copper ingots.

Bass went from this project—which he thought would be his only experience with excavating a shipwreck—to another, also off the Turkish coast. This one was Byzantine. Then there was a Roman wreck. At this rate, he was becoming an underwater archaeologist.

He and his team created methodology as they worked. This included the design and construction of a support that would travel on a grid and stand on the sea bottom as a base for the photographers. They made a bubble called "the telephone booth," in which divers could remove their masks and speak by telephone with others on the dive boat. They developed a system of three-dimensional photographic mapping. Since the first step in exploring wrecks was finding them, they started using sonar to locate wrecks.

Bass founded the Institute for Nautical Archaeology (INA) in 1973, in large part to give weight to the claims that this was indeed a scientific discipline, not simply underwater treasure-hunting. The institute, Bass, and the research team moved to Texas A&M in 1976, and are still there; a Turkish base of INA is located in Bodrum.

In 1984, Bass went to look at an unexplored wreck that had been located two years earlier. It lay on a slope 140–170 feet down, and was dated as a ship of the 14th or 13th century B.C.E.—the world's oldest shipwreck. Debris was scattered everywhere—copper ingots, amphorae (storage jars), and other objects. The excavation and preservation of this ship, a major project, received funds from the National Geographic Society and the National Science Foundation, among others. Bass and his crew worked on the site for 11 seasons. The lengthy expedition was necessary because working underwater presents significant problems. Work periods at that depth can be only about 20 minutes long before nitrogen narcosis makes cognition impossible. Then, on the way up out of the water, decompression of the divers takes hours.

Other objects were found in the wreck. Among them were glass disks, and metal ingots of copper and tin in the 10:1 ratio used for making bronze. The glass was similar to that known in the 18th Egyptian dynasty. Ivory in the form of hippopotamus teeth and segments of elephant tusk, aromatic resin, and the black wood ebony,

known in ancient Egypt, were also part of the cargo. A gold chalice, a scarab, and other small gold objects were the sensational finds. Bass believes this cargo was a present of luxury goods designed to cement a diplomatic treaty of its day. He ties a number of references in the *Iliad* to materials found in this wreck. According to Bass's theory, the poet Homer lived earlier than his accepted dates in the eighth century B.C.E. and his epic poem described contemporary events.

The work of INA continues. There are still shipwrecks to be explored, and the techniques developed are now accepted as archaeology. The remains of one of the ships that Bass explored are exhibited in the Bodrum Museum of Underwater Archaeology. The museum is part of the large Bodrum Museum now housed in the fortress built by the Crusader order, the Knights of St. John. It is on the site of ancient Halicarnassus.

⊠ **Bede**
(672?–735)
English
Natural Scientist

The English scholar-scientist monk, Bede, never traveled more than 50 miles from his monastery. He was sent as a student to Benedict Biscop, the abbot, when he was seven years old and stayed for the rest of his life. Biscop surrounded himself with the best scholars he could attract and created a brilliant court and school in northeast England, centered on the two monasteries he founded in Wearmouth and Jarrow. The latter was Bede's home. Most of Bede's writings are concerned with monastic life and ecclesiastical subjects. One book, however, is an encyclopedia of earlier works on solar and lunar cycles. A major question of his day was the determination of the date for Easter. The establishment of a good, working calendar was his scholarly endeavor.

Bede's contribution to ocean sciences was a detailed description and tables for tides. This is

An artist's sketch of the "Venerable" Bede. *(Hulton/ Archive/Getty Images)*

no surprise: he lived near Lindisfarne, which at high tide is an island in the North Sea, and is the end of a peninsula at low tide. He knew, as did everyone who tried to establish tide tables, that tides and lunar cycles went together. Bede came close to a working concept of longitude as he noted the progression of high tide westward.

By the 11th century, scholarship in England seriously declined. This was due to internal wars and Norse raids, and Bede's reputation as a scientist declined as well. He was known only as a geographer; the significance of his work on tides and currents was totally overlooked. His works were preserved because of his efforts in the creation of the Western calendar that was later refined by Pope Gregory.

⊠   **Beebe, Charles William**
(1877–1962)
American
Naturalist

The man best remembered for his spectacular, record-setting undersea descent is Will Beebe, born in Brooklyn, New York, on July 29, 1877, and raised in East Orange, New Jersey. He was the only child of Charles, a paper dealer, and Henrietta. The family's interest in natural history made them regular visitors to the American Museum of Natural History in New York City— so much so that the teen-aged Will was well known to the staff. The president of the museum, Henry Fairfield Osborn, befriended him and encouraged his interest in birds. Osborn was also a professor at Columbia University and the first president of the New York Zoological Society. It was the zoological society that created the New York Zoological Park, commonly known as the Bronx Zoo. It has since changed its name to the Wildlife Conservative Center. Beebe went to Columbia, never graduated, but went on to the zoo as Assistant Curator of Birds in 1899. The post of Curator of Birds became his in 1902. This post took him on an extensive trip to catalog and study pheasants in the wild. After a 17-month trip, he published the first volume of *A Monograph of the Pheasants* in 1918. Then he turned his animal research to wildlife in tropical forests and continued writing and research in the field.

Beebe's career in oceanography began in the 1920s, when he learned to dive using helmets and hard suits. He went on several expeditions at sea on the *Arcturus*, where he studied the animals that had been dredged up. Since he wished to see these creatures alive and in their natural habitats, he enlisted Otis Barton in this endeavor. Barton was a trained engineer and had the money to fund his own research. They began their joint venture in 1928 and built a diving chamber from which they could photograph the life in the sea around them. In August of 1932, Barton and

Beebe, on their third dive in the bathysphere, descended to a depth of over 900 m (3,028 ft). In 1949, Barton, in a redesigned vessel, dove 1260 m (4,500 ft) below the surface.

Beebe was a leader in the zoological world because he concentrated on the biome of the animals he was studying. His emphasis on the animals' natural setting led him to maintain a tropical research station in Guyana, then in Bermuda, the site for his bathyscaphe dives, and after World War II, at Simla, his estate in Trinidad. His encouragement and aid to RACHEL CARSON enabled her to continue her writing, producing the best-seller *The Sea Around Us*.

Will Beebe retired from his position as director of the Department of Tropical Research in 1952. He continued his scientific interests and writing, and died on June 4, 1962. Simla, his estate, was bequeathed to the New York Zoological Society as one of its research preserves.

Naturalist and explorer Dr. Charles William Beebe and members of his party stand by his bathysphere on the SS *Monarch* on December 7, 1932, as they return to New York from the coast of Bermuda, where they studied ocean depths. *(Hulton/Archive/Getty Images)*

⊠ **Bellingsgauzen, Faddei Faddeevich**
(Bellingshausen, Fabian Gottlieb von)
(1778–1852)
Russian
Explorer, Oceanographer

Alexander I, czar of Russia, chose Bellingsgauzen to lead an expedition to find the purported southern continent, Terra Australis. This had been one of the aims of JAMES COOK's expeditions, and after Cook's death in 1779, a number of attempts to continue his work were funded by seagoing nations. When Bellingsgauzen was given his orders, he studied Cook's papers and considered him a mentor and personal hero.

The future admiral began his career when he went to sea as a cadet at the age of 10. After graduation from the Naval Academy at Kronshtadt in 1796, he sailed almost constantly in the Baltic Sea. When the czar sent him to find Terra Australis in 1819, Bellingsgauzen was ready for the challenge. The *Vostok,* his flagship, was accompanied by the *Mirnyi,* a transport vessel. Their first stop was Portsmouth, where Bellingsgauzen traveled from the port to London. There, Joseph Banks, head of the Royal Society and naturalist on Cook's second voyage, supplied the Russians with books, maps, and charts.

Leaving England in September, they were at South Georgia, an island near the Falklands, by the year's end. The Russians circumnavigated the South Sandwich Islands, using icebergs as fixed points and a fresh water supply. The ships crossed the Antarctic Circle on January 26, 1820, and came within 20 miles of the Antarctic continent. Whether Bellingsgauzen was the first to see the mainland is in dispute. The English claim it was Edward Bransfield, a naval captain, and the Americans claim it was Nathaniel Palmer, a seal-hunting captain. Surviving a terrible storm, the two Russian ships made port at Sydney, Australia. After rest and repair, they set off to explore the Pacific, and on returning to Australia, were confronted by the news that an English captain had found a group of islands that he named the South Shetlands, and which he believed were part of the Antarctic. Bellingsgauzen went to look for them, leaving in November of 1820.

Going again south of the Antarctic Circle, the two ships encountered several storms that pushed them north again and again. On January 21, 1821, Bellingsgauzen believed he spotted land, naming it the Alexander coast in honor of the czar. The sighting was not coast but Alexander Island. Continuing south, the Russians encountered an American sealer, Captain Palmer, of the *Hero.* Their accounts of this meeting differ in that each claims to have been the first to see Antarctica. Bellingsgauzen named the southern Antarctic Peninsula Palmer Land, in honor of his rival.

The return voyage was uneventful, and the ships arrived in Kronshtadt in August 1821. Russia promptly lost interest in the expedition. Bellingsgauzen's papers were unpublished for more than 10 years after his return. His career in the navy continued; he achieved admiralty rank and later served as governor of Kronshtadt, where he died on January 13, 1852. Russia's interest in the southernmost continent was rekindled in the mid 20th century. As a part of a number of international efforts near the South Pole, a soviet scientific base was established in the Antarctic during the International Geophysical Year (1957–58).

⊠ **Bernoulli, Daniel**
(1700–1782)
Swiss
Mathematician

Daniel Bernoulli was born on February 8, 1700, in Groningen, Netherlands, where his father, Johann, was the professor of mathematics. The family was one of noted mathematicians. Both Daniel's brother Nicholaus II and his uncle Jacob were well known; Daniel's mother, Dorothea Falkner, was of a patrician family. The Bernoulli family returned to their native Basel,

Switzerland, in 1705, when Johann was chosen to fill the university's mathematics chair that had been vacated by the death of his brother Jacob. Johann forbade his son to study mathematics because there was no money in it. After Daniel's completion of his master's degree in philosophy in 1716, he was apprenticed to a merchant. After that attempt at a commercial career failed, Johann sent Daniel back to the university, this time to study medicine. At the same time, he was taught calculus at home. In the fashion of university education of his day, scholars traveled to hear specific professors lecture in their disciplines. Thus, Bernoulli spent a year in Heidelberg (1718) and another in Strasbourg (1719) and finally finished his medical studies in 1720. His father continued to teach him mathematics at home and by letter, and he worked on the theories of kinetic energy and conservation of energy, applying these concepts to medicine. The doctoral dissertation Bernoulli presented was on the mechanics of breathing. This subject area might now be called biophysics, the physics such as pressure, fluid flow, and torque, that affect living things.

After applying unsuccessfully for a university post in Basel, Bernoulli went to Venice to study practical medicine. His first published work, *Mathematical Exercises*, was produced there in 1724. In it he discusses the flow of blood and blood pressure, and these studies gave him an interest in fluid flow that would continue in other works, significantly in naval design. For example, while in Venice he designed an hourglass to be used at sea. This was important because the flow of sand was constant even on a ship pitching in heavy seas. This invention won a prize from the Paris Academy of Sciences. The attention generated by the prize, combined with the publication of the book, won him an appointment at St. Petersburg, Russia. On his journey there in 1725, he was accompanied by his older brother Nicholaus.

Within months of their arrival, Nicholaus died, and Daniel wished to return to Basel. Since Johann wanted him to stay, he sent one of his

Daniel Bernoulli *(E. F. Smith Collection, Rare Book & Manuscript Library, University of Pennsylvania)*

own students, Leonard Euler, to St. Petersburg to work with Daniel. This association proved to be a fruitful one. Euler arrived in 1727, and Daniel stayed in St. Petersburg until 1733. During that time, he worked on vibrating systems. In this study, he analyzed the nodes and frequencies of vibrating strings, which is what create different sounds when a violin is played. This was the beginning of wave theory, an important component in the study of such other fields as the propagation of both sound and light. His most important work on hydrodynamics was also begun at this time. Bernoulli coined the term for the field in his title *Hydrodynamica*. The work is, in part, an exacting and correct analysis of water flowing from a container. It also discusses the mathematics of pumps and machines that raise water. Another chapter deals with the kinetic theory of gases. In this section, Bernoulli mapped out the equations of state that were fully described by Johann Diderik van der Waals, the 19th-century physicist and Nobel prize winner, a century later. The theory explains the pressure of a gas on the walls of its container as an expression of kinetic energy. Bernoulli's Law or Principle is also a significant part of this book. It describes the energy of a stationary mass of fluid (gas or liquid) in terms of its gravitational potential energy, and describes what will happen if the fluid is moving, its kinetic energy. The way a fluid moves over and around obstacles is an essential part of the design of boat hulls and airplane wings.

Bernoulli continued to apply for university positions that would take him back to Basel, and in 1734 he finally was successful; he was to teach botany. That year, both father and son submitted entries for the Grand Prize of the Paris Academy. Johann's work was on dynamics, and Daniel's was on astronomy. Unhappily, their entries were named joint winners. This so infuriated Johann that the son was banned from his father's house, a break that was never healed. Daniel Bernoulli continued a correspondence with Euler, and he added material to *Hydrodynamica* explaining jets of fluid and the propulsion of ships. He also involved himself in practical problems. One such application received the 1737 prize of the Paris Academy. It was awarded for the design of a ship's anchor to both Daniel Bernoulli and Giovanni Poleni, the shipwright who executed Bernoulli's design.

Hydrodynamica was finally published in 1738. In 1739 Johann published his great work, *Hydraulica*, predated to 1732 to look as if the son based his work on that of the father, when the reverse was more likely the case. With the opening of a post at the university, Daniel switched from botany to physiology, and in 1750, was appointed to the chair of physics that he kept until 1776. While his experiments in physics remained unpublished, he predicted several laws that were later verified. For example, Charles A. Coulomb (1736–1806) defined the interaction between electrical and magnetic force. This states that in

a vacuum, if two electrical charges are distant from one another, the force between them is represented by the product of the charges divided by the distance separating them. Bernoulli had demonstrated this, but not written it for publication. During this period, his lectures in physics were popular enough to be attended by the intellectuals of the day.

Still in Basel, Bernoulli won the Paris Academy's prize 10 times for astronomical and nautical topics. Euler was coauthor with him in 1740 when they won for an explanation of Newton's theory of tides. His other winning contributions were in 1743 and 1746 on magnetism, in 1747 for a method of determining time at sea, in 1751 for a treatise on ocean currents, in 1753 for effects of forces on ships, and in 1757 for a discussion of methods to reduce pitching and tossing of ships in rough seas.

Bernoulli, in spite of his father's animosity, did manage to work successfully with other family members. He was much honored in his own time and elected as a member of most of the leading scientific societies of Europe. He died in his beloved Basel on March 17, 1782.

⊠ Bjerknes, Jakob Aall Bonnevie
(1897–1975)
Norwegian
Meteorologist

Jakob Bjerknes was born in Stockholm, the second of four sons, to a family with prestigious scientific backgrounds: his father, VILHELM BJERKNES, was declared to be "the father of modern meteorology"; Honoria Bjerknes, his mother, studied natural science; and Kristine Bonnevie, her sister, was the first professor in Norway of zoology and embryology. Bjerknes was educated in Oslo and later joined his father in Leipzig, Germany, in 1912.

When they returned to Norway in 1917, Jakob became the head of the weather forecasting station in Bergen. Bjerknes's weather studies had

Jakob Bjerknes *(American Geophysical Union, courtesy AIP Emilio Segrè Visual Archives)*

in the stratosphere. The upper air layers are determining factors in the weather in the lower atmosphere.

The German invasion of Norway in 1940 occurred while Jakob Bjerknes was in California, where he was introducing Norwegian weather forecasting programs to American meteorologists. The University of California at Los Angeles (UCLA) offered him a faculty position, and he accepted it. Together with Jorgen Holmboe, a former assistant of his father's, Jakob developed the meteorology department at UCLA and trained military meteorologists. After the war ended, there was ongoing cooperation and joint projects between American meteorologists and "the Norwegian mob"—BJØRN HELLAND-HANSEN and HARALD SVERDRUP—at Scripps. Jakob became Jacob, then Jack, and, as an American citizen, continued his research on the interaction between ocean and air circulation.

Other projects that Bjerknes was involved with include the tracking and study of El Niño, the jet stream, and the coordination of all air and ocean current studies. He was still working on these projects when he died in Los Angeles on July 7, 1975.

really begun in 1918 in Leipzig when he realized that cyclones are weather fronts. They are three-dimensional surfaces of discontinuity between air masses of different origin and characteristics. In Leipzig, he had had available the doctoral work of Herbert Petzold, one of his father's students, who had studied line squalls. These rapidly moving air masses are a major hazard for pilots. Bjerknes extended this work to produce a detailed mathematical explanation of what is actually happening in a cyclonic storm. Evolving instrumentation transformed this field, and he continued in it throughout his life.

In 1931, he was named professor of meteorology at the Bergen Museum. His research interest was the relationship between weather phenomena in the troposphere and wave patterns

⊠ Bjerknes, Vilhelm Frimann Koren
(1862–1951)
Norwegian
Meteorologist

Acclaimed by other meteorologists as "the father of modern meteorology," Vilhelm Bjerknes was born on March 14, 1862, in Christiania, later Oslo, when Norway was a part of Sweden. His father was Carl Anton Bjerknes, a professor of mathematics at the University of Christiania, who devoted his life to the mathematical description of the movement of solids through frictionless fluids, a branch of hydrodynamics. Vilhelm Bjerknes was educated at home and at the university, receiving his M.A. in mathematics in 1888. He was anxious to make his own scientific start, not just to continue working on his father's theoretical projects. Since his mathematician father was becoming increasingly reclusive, the young man wanted other exposure. He applied for and was awarded a state scholarship in 1890. This grant enabled him to travel, and he promptly left for Paris.

Once in Paris, he attended Jules-Henri Poincaré's lectures on electrodynamics and then went on to Germany to continue this study with Heinrich Hertz, a leader in the field. He stayed with Hertz first as a student, then a coauthor, incorporating Poincaré's work to prove Hertz's theories on electrical resonance. Hertz and Bjerknes maintained a friendship as well, and this continued after the latter returned to Norway in 1892.

Bjerknes finished his Ph.D. in physics shortly after his return to Norway. He was appointed as lecturer in applied mechanics at the Engineering School in Stockholm, Sweden, the next year. There, he began to develop his theories on hydrodynamic circulation, partly a continuation of his father's work. He reexamined the work of Hermann Helmholtz and William Thomson, Lord Kelvin, on the velocities of circulation vortices, applying his findings generally to the motion of fluid in oceans and wind currents. When he returned again to Norway in 1907, it was as professor of applied mechanics and mathematical physics at the University of Christiania, where he worked with Johan Sandström, Theador Hesselberg, Olaf Devik, and HARALD SVERDRUP.

Meanwhile, a lecture presentation of the study of hydrodynamics and thermodynamic change, as related to movements in water and air, took Bjerknes to the United States in 1905. There, Bjerknes received a Carnegie Foundation grant for a professorship in geophysics and a geophysical institute. This venture was hardly begun in Oslo when an offer of space and support occasioned its move to Leipzig, Germany, in 1912.

FRIDTJOF NANSEN offered Bjerknes a professorship at the Bergen (Norway) Museum—there was no university there yet—and the chance to bring his geophysical institute back to Norway. He accepted, and his tenure as director lasted from 1917 to 1926. This was the most productive phase of his career. The Norwegian weather service was standardized and enlarged due to his efforts as director of the institute. (He launched the studies of cyclones, air masses, and weather fronts.)

In 1921, he published his classic work, *On the Dynamics of the Circular Vortex with Applications to the Atmosphere and to Atmospheric Vortex and Wave Motion.* With his son JAKOB BJERKNES, and Halvor Solberg, Bjerknes published *Physikalische Hydrodynanmik mit Anvending auf die dynamische Meteorologie* (Physical hydrodynamics with applications to dynamic meteorology) in 1933. He coauthored the first two volumes of *Dynamic Meteorology and Hydrography.* The last volume of this work appeared in 1951. In addition to his contributions to meteorology, Bjerknes continued the work with Helmholtz and was involved in the early development of wireless telegraphy. He died on April 9, 1951.

Bory de Saint-Vincent, Jean-Baptiste-Geneviève-Marcellin
(1778–1846)
French
Natural Scientist

Bory was born in Agen in the southwest of France on July 6, 1778, into a prestigious family that pursued the French tradition of amateur natural historians. Because his education was interrupted by the French Revolution from which his father was a fugitive, his maternal uncle became his mentor in the study of natural history. As an adolescent, Bory submitted his first scientific publication on *Conferva* and *Byssus*, two marine organisms, to the Academy of Bordeaux. In 1797, he joined the army and was posted to Belle-Ile on the Atlantic coast. With the help of BERNARD LACÉPÈDE, then working at the Museum of Natural History in Paris, Bory was appointed as a naturalist on the

Bougainville, Louis-Antoine, comte de   23

French Pacific expedition of 1800–04 under the command of Nicolas Baudin.

The ship sailed south past Madeira, the Cape Verde Islands, and the Canaries. By the time they arrived in l'Île de France (now Mauritius), a near-mutiny erupted and several members of the crew and some of the scientists, Bory among them, left the ship. He remained on the island and also explored Réunion, a nearby island. There, he collected many kinds of natural specimens; plants were his first love, but he did not neglect the local animals, and rocks. He also observed spectacular phenomena. When a major volcanic eruption occurred, he described it thoroughly. A passing German vessel gave him passage to France. It stopped at St. Helena on the journey, where Bory caught a spectacular butterfly. He presented an accurate map of the island and the butterfly to Napoléon on his return.

Marriage in 1802 and continuation of his interest in natural history kept him busy. He continued in the army, working as a cartographer, and maintained his career as a naturalist. The publication of a two-volume work on his travels with the Baudin expedition was a success.

Politics nearly ruined his scientific as well as his military career. With the first exile of Napoléon in 1814, Bory swore to uphold the king, Louis XVIII. With the return of Napoléon, Bory again supported the emperor. After the battle of Waterloo, where Napoléon was defeated, Bory was banished. He spent years on the run, still collecting plants and rocks. By 1819, he was pardoned and returned to Paris penniless, having been stripped of his army pension. Going back to his beloved plants, he worked diligently as a botanist from 1820–23. Many of the entries describing algae in the *Dictionarie classique d'histoire naturelle* (Classic dictionary of natural history), published in parts by the Musée d'Histoire Naturelle from 1822–31 were written by Bory. In addition to these, he described diatoms and sponges, and speculated on the possibility of a three-kingdom division of living organisms.

Bory's extravagant style of living resulted in another personal disaster. Imprisoned for debt in 1825, he continued to work on his algae in prison. The specimens he studied were brought back to France by JULES DUMONT D'URVILLE, the first officer on the *Coquille*, which sailed around the globe under the command of LOUIS-ISIDORE DUPERREY. Bory named specimen genera for Dumont d'Urville and for the ship's botanist, René P. Lesson: *Durvillaea* and *Lessonia*. He was released from prison in 1828, when his daughter's fiancé paid his debts.

Another revolution in 1830 brought a new king, Louis-Philippe. The edict of 1815 that had deprived Bory of his army pension was rescinded and, as a colonel, he received all his back pay. In 1832, he was elected to the Academy of Sciences, and in 1839, nominated to the scientific commission that was sent to study the natural history of Algeria. He had a wonderful time, spent his entire fortune, and returned to Paris in 1842. He died suddenly, on December 12, 1846, leaving a mountain of debt, and a fantastic herbarium that was purchased by a private buyer who gave it to the nation.

In addition to his scientific work, Bory wrote two comedies, fables, and verse. He made splendid maps of every place he visited with the Grande Armée, and when in exile, included the flora and fauna of the caves near Maastricht, in the Netherlands, where he had lived in hiding. He was a truly courageous, funny, capable man who enjoyed being a scientist and a *bon vivant*.

⊠ **Bougainville, Louis-Antoine, comte de**
(1729–1811)
French
Explorer, Naturalist

Born in Paris on November 11, 1729, the son of Pierre-Yves Bougainville, Louis-Antoine escaped his father's profession—he was a notary—by

French navigator Louis-Antoine de Bougainville *(Roger Violett/Getty Images)*

joining the army. The military brought him to North America, where he fought at Quebec under the command of Louis-Joseph de Montcalm, the French field marshal. In 1763, he left the army and joined the navy as a captain. Bougainville founded a colony on the Falkland Islands in 1764. The French navy then commissioned him to circumnavigate the globe in 1766.

At the end of 1766, he departed from Nantes on the voyage that made him famous. The first assignment was a stop at the Falkland Islands, where he officially passed the islands to Spain.

The second assignment was meeting the supply ship in Rio de Janeiro, capital of the Portuguese colony of Brazil. Philibert Commerson, the botanist on that vessel, had collected a plant near Rio, then unknown to French botany. The purple red bracts made this climber an attractive plant, and Commerson named it for Bougainville. Together, the two ships sailed south, through the Straits of Magellan and into the Pacific on a reconnaissance mission. The Tuamoto and Solomon Island groups were found and

charted. In the Moluccas, Bougainville accurately described marsupials. The systematic astronomical observations that the expedition made were incorporated in chart making and determination of longitude. Bougainville's expedition returned to France in 1769.

Bougainville's first publication, in 1752, was a treatise on integral calculus, written before his army service. His *Voyage autour du monde* (Voyage around the world) was published in 1771, shortly after his return. It was a popular success and continued to be printed for years. It contains his famous quote, which is a great example of his attitude toward practical work: "Geography is a science of facts: one cannot speculate from an armchair without the risk of making mistakes which are often corrected only at the expense of the sailors."

This remarkable man managed to live in Paris throughout all the political upheavals of the end of 18th century. He was elected as an associate of the Académie des Sciences and a member of the Legion of Honor, was ennobled as a count in the First Empire, and held the office of senator. Much honored in life, he died on August 31, 1811, and is buried in the Panthéon in Paris.

Bowditch, Nathaniel
(1773–1838)
American
Navigator, Astronomer

This native of Salem, Massachusetts, was a self-taught polymath. He achieved great success as the author of *The New American Practical Navigator,* a book that was first published in 1802 and is still being printed. This unusual volume was the first reliable handbook of navigation tables. It also presents information on winds and currents, and clear instructions for instrument calculation of location at sea. One of the things that makes this so remarkable is that Bowditch worked alone and taught himself algebra, calculus, geometry, trigonometry, and several foreign languages.

Born on March 26, 1773, Bowditch was the fourth of seven children of a shipmaster, Habakkuk Bowditch, who had fallen on hard times and had become a cooper. Schooling ended for the boy when he went to work for his father at age 10. Two years later, he was apprenticed to a chandler. That is when he began his solitary study of mathematics and began to buy sextants, gradually amassing what became a significant historical collection of the instruments. A lucky chance brought him in contact with part of a privateer's booty, the library of Dr. Richard Kirwan, an Irish chemist and naturalist. This included the works of the Bernoullis, both DANIEL BERNOULLI and his brother Jakob, Isaac Newton, and Robert Boyle, as well as the *Encylopaedia Britannica*. Reverend Joseph Willard, a Harvard-trained man, persuaded a group of prominent Salem citizens to buy the library for the town's Philosophical Society. Another minister, William Bentley, who had recognized the boy's talent, allowed him to borrow the books. Some were in Latin, others in French. He taught himself both languages by comparing Bibles in those tongues with his own English Bible.

Using Newton's work led to his first publication when he was 17. He found an error in Newton's *Principia* and wrote an article explaining it. Before publishing this, he wrote to a Harvard professor asking that his work be checked. The professor, after first ignoring the letter, finally answered. The Salem unknown was correct— Newton had made a mistake.

Bowditch first went to sea in 1795 as a clerk; over time he was promoted to supercargo, and finally master. Salem was a rich port with dealings in the Far East, which meant that Bowditch had opportunities to travel to the East Indies and China several times. While on these long voyages, he began to study the navigation tables available. Although John Harrison had invented an accurate chronometer for the determination of longitude in 1761, this expensive piece of equipment was not widely accessible. Navigation was largely a matter of dead reckoning, gauging distances from one fixed point to another. Using the tables he could find, Bowditch set about correcting the roughly 8,000 errors he found in them. This grew into his great work, *The Practical Navigator*. The major innovation in this work was the determination of latitude and longitude using lunar observation. The resulting calculation produced a position that was, at worst, 30 miles off, far better than any other system then used.

Harvard College awarded Bowditch an honorary M.A. within months of the publication of *The Practical Navigator* and a doctorate several years later. Bowditch revised and expanded this publication throughout his life and was succeeded in this work by his son Jonathan, until 1868, when the navy's Hydrographic Office continued it.

American astronomer and mathematician Nathaniel Bowditch; engraving by J. Cross after a drawing by J. B. Longacre of a bust by Frazee (ca. 1830). *(Hulton/Archive/Getty Images)*

While president of the Essex Fire and Marine Insurance Company from 1803 to 1823, Bowditch spent three years working on a survey of the harbors of Beverly, Salem, Marblehead, and Manchester, Massachusetts. Although President Thomas Jefferson began to organize an agency to do this work, it was years before coastal surveys became a priority with the federal government. The position of surveyor enabled Bowditch to finance the publication of his translation of Pierre Laplace's *Méchanique Céleste* (Celestial mechanics). This project occupied him from 1814 until the end of his life. The first three volumes appeared in his lifetime, a fourth was published posthumously, and a fifth volume was based on his notes. These volumes were not just translations but included commentaries and corrections. This monumental redaction made it possible for English speakers to know the classic Laplace work that applies Newtonian mechanics to astronomy. The original was published in Paris over a 10-year period, from 1795 to 1805.

Never comfortable with speaking in public, Bowditch turned down offers of professorships at Harvard, the University of Virginia, and the U.S. Military Academy. He did, however, accept membership in the Harvard Corporation, the organization that runs Harvard College. One of his activities in that enterprise was the reorganization of the finances of the college. This self-taught wonder was elected to every major scientific society in the United States and in Europe. He died in Boston on March 16, 1838, and is buried in Salem.

⊠ **Bowie, William**
(1872–1940)
American
Geodesist

William Bowie was born near Annapolis, Maryland, on May 6, 1872. He attended Saint John's College there before transferring to Trinity Col-

William Bowie *(American Geophysical Union, courtesy AIP Emilio Segrè Visual Archives)*

lege in Hartford, Connecticut, in 1893. After earning a liberal arts degree, he pursued engineering at Lehigh University in Bethlehem, Pennsylvania, where he received a degree in 1895. In that year, he joined the U.S. Coast and Geodetic Survey and remained there throughout his career, returning to Hartford for an M.A. in physics in 1907. His duties included careful geodetic, topographic, and hydrographic measurements in the continental United States and also in Alaska, Puerto Rico, and the Philippines. He was made chief of the Geodesy Division in 1909, retiring in 1936.

Continuing the work of his predecessor, John Hayford, in the U.S. Coast and Geodetic Survey was one of Bowie's major aims. He supported

Hayford's view of Earth's shape as a geoid, and his concept of isostasy. Hayford had theorized that the Earth was not a sphere but an ellipsoid in which the solid rock and the liquid magma below it were in equilibrium. If some solid rock emerged, as in a volcanic eruption, somewhere on Earth there had to be a subsidence. This idea is still part of the history of theoretical geology since, when Hayford proposed it, he could not foresee the later work of W. MAURICE EWING, HARRY HESS, and others that demonstrated the movement of the crustal plates that form the Earth's surface. Isostasy was a concept Hayford created that explained the state of the Earth, given the information he had then. He said elevations in the Earth's crust are balanced by mass below the surface of the semi-liquid mantle; thus a state of equilibrium of pressure exists independent of geographic position. As a supporter of FELIX VENING MEINESZ's work on gravity, Bowie advocated the expedition that recorded gravity at sea. Vening Meinesz, a Dutch researcher, was then active. He was making careful measurements of gravity at different places on the oceanic surface.

In his position as head of the Survey, Bowie urged the various federal agencies that produced maps to coordinate and standardize these efforts. His book, *Isostasy* (1927), became a classic in the field. His almost 400 publications were compiled as part of a memoir published by the National Academy of Sciences after Bowie's death on August 28, 1940.

Several universities in the United States and abroad awarded Bowie honors, and he also received recognition from the National Academy of Science, and other academies of science abroad. Bowie was instrumental in forming the International Geodetic Association after World War I and in the creation of international organizations for geodesy and geophysics. He was the first president of the American Geophysical Union and recipient of its medal for fundamental research and unselfish cooperation in research.

Thus, he crossed both scientific and political boundaries with a diplomat's skill and made possible the foundation of multinational research organizations.

⊠ Brookes, William Keith
(1848–1908)
American
Biologist, Environmentalist

William Brookes was born in Cleveland, Ohio, on March 25, 1848, coming east to Williamstown, Massachusetts, to attend Williams College. There he concentrated on biology and received a degree in 1870. A meeting with LOUIS AGASSIZ in 1873 at the Anderson School of Natural History so interested Brookes in marine animals that he was determined to specialize in their study.

His mentor was ALEXANDER AGASSIZ. In 1875, Brookes was the recipient of the third Ph.D. that Harvard awarded. An appointment as associate in biology in Baltimore, Maryland, at the brand new Johns Hopkins University followed. There, he worked with H. Newall Martin. Martin had a notable mentor in Thomas Huxley, CHARLES DARWIN's great defender and supporter in the controversy surrounding the introduction of the Darwinian theory of evolution. Together they developed a curriculum for graduate education, basing their course outline on the German model.

By 1894, Brookes was chair of the biology department at Johns Hopkins, a position he held until his death. The body of research he produced was based on the data taken by the Chesapeake Zoology Laboratory, a movable experimental station that operated from 1878 to 1906. This laboratory moved up and down the Atlantic coast and to the West Indies. Study of organisms in their environment was Brookes's specialty. In his writings, he tried to bring together GEORGES CUVIER's ideas of function of organs and structure with the Darwinian concept of the history of change. To Brookes, "survival of the fittest" meant continuation of usable, versatile structures. Brookes conducted studies of the development of the embryos of tunicates, coelenterates, and crustaceans that are now considered classic. While his works on morphology have not withstood the test of further research, the students Brookes trained, notably E. B. WILSON, T. H. Morgan, E. G. CONKLIN, and R. G. Harrison, went on to develop the fields of cytology, embryology, and genetics. The environment of the Chesapeake Bay area was extremely important to Brookes. However, his warnings of commercial overuse of the bay area's shellfish were ignored.

Always in fragile health, Brookes died of heart failure on November 12, 1908.

⊠   Buffon, Georges-Louis LeClerc, comte de
(1707–1788)
French
Biologist, Natural Scientist

Born in Montbard, central France, on September 7, 1707, Georges-Louis LeClerc was originally from a bourgeois family. His father, François LeClerc, married well; Anne-Christine Marlin was an heiress, and he became rich enough to buy a title. The future comte de Buffon's early education was at the Jesuit College in Dijon, where he drifted away from his father's legal profession to mathematics and then to Angers for further study. He left Angers because of his involvement in a duel and accompanied a young English gentleman on his three-year-long grand tour of southern France and Italy. After returning in 1732, he became a regular at the Parisian science salons and worked on naval assignments determining the strength of materials important in shipbuilding.

Meanwhile, he devoted considerable time to his business affairs, managing his growing estate, and doing research; the business supported the re-

search. During the 1730s, Buffon published his translations of Stephen Hales's *Vegetable Statiks* and Isaac Newton's *Methods of Fluxions and Infinite Series* (calculus). Based on these works, he rose in rank as a member of the Académie des sciences.

In 1739, Buffon succeeded Charles-François DuFey as *intendent* (curator) of the Jardin du roi (the king's botanical garden). This was not just an ornamental garden but a scientific research center, and the post was a significant one. For most of his time there, Buffon worked on his 36-volume *Histoire Naturelle* (Natural history). This work appeared in sections from 1737 to 1752. Buffon's ideas about relationships between organisms changed as he worked on this subject and became even more elaborate by the time the supplement appeared in 1774–77. By then, Buffon was a major figure in the scientific world and had been granted the title of count by Louis XV. He was a member of the Académie royale des sciences, the Académie française, the Royal Society of London, and the academies of Berlin and St. Petersburg. This international celebrity continued in his post at the Jardin du roi until his death on April 16, 1788.

Buffon attempted to separate science from religious and philosophical concepts and to base his study on observation alone. He wrestled with the ideas of Carolus Linnaeus, the Swedish botanist known for his scheme of classification of all organisms, living and extinct. Buffon tried to bring systematization to biological family-order-species concepts. The notion that every living thing could be arranged in a taxonomic structure was a daring new thought at this time, and initially he resisted it. His knowledge of animal construction was phenomenal, and he is credited with having delineated the science of paleontology. Buffon was a forerunner of JEAN-BAPTISTE LAMARCK in his belief that inheritable changes could be caused by environment, climate, or food. He believed in gradual transformation as a mechanism in evolution and heredity. Observa-

Georges-Louis Buffon *(E. F. Smith Collection, Rare Book & Manuscript Library, University of Pennsylvania)*

tion forced him to acknowledge variability in related animals, although he had no means to explain heredity.

In his *Memoirs,* Buffon wrote on a wide variety of subjects ranging from mathematics, optics, and astronomy to forestry and plant physics. In the *Épochs de la nature* (Epochs of nature), using his considerable knowledge of fossils and comparative anatomy, he estimated the age of the Earth at about 75,000 years, when the accepted religious figure was close to 6,000 years. Further work on rates of sedimentation led him to revise that figure to 3 million years. Afraid of being misunderstood, he did not publish his theory. The comte de Buffon is important in the history of biology because he was a great collector and systematizer. His position in the Jardin du roi (renamed Jardin des plantes

after the start of the French Revolution of 1789) elevated and maintained that institution, which was central to scientific development and essential for the education of generations of scientists.

⊠ Bush, Katherine
(1855–1937)
American
Biologist

Born to William Bush and Eliza Ann (Clark), Katherine Bush was raised in Scranton, Pennsylvania. She first appeared on the scientific scene as an assistant to ADDISON VERRILL, who was the professor of biology at Yale University and a primary researcher for the U.S. Fish Commission. Bush worked as his assistant in the Peabody Museum of Natural History, Yale University's collection, for more than six years. They were both researchers and taxonomists for the U.S. Commission for Fish and Fisheries.

It became obvious that Bush could have an independent research career, and she enrolled in Yale's Sheffield Scientific School in 1885, where she studied for the equivalent of a bachelor's degree. At the time, Yale was not awarding the B.S. degree, and scientific disciplines were not part of the Yale curriculum. Bush graduated from Sheffield with a degree in biology and continued on at Yale to earn a Ph.D. in zoology in 1901. She was the first American woman to do so. She then went on to further study in biology from 1901 to 1904 and again from 1908 to 1909.

Bush's first scientific interests were the Mollusca and Echinodermata. She had much experience with mollusks—shelled invertebrates such as clams, oysters, and snails—and echinoderms, the starfish, sea urchins, and brittle stars, from her days as a sorter in the U.S. Fish Commission. Her publications based on this work are "Catalogue of the Mollusca and Echinodermata . . . Labrador," in *The Proceedings of the U.S. National Museum 1883 and 1884;* "Report on Mollusca Dredged by the

Blake in 1880," in *The Bulletin of the Museum of Comparative Zoology* (1893); and "Revision of Deep Water Mollusca of the Atlantic Coast of North America: Bivalvia," in *The Proceedings of the U.S. National Museum* (1898). The last was coauthored by A. E. Verrill. These publications are the standard works on the organisms described.

Later in her career, she studied the specimens brought back by the Harriman expeditions to Alaskan waters (1894–99). The organisms of interest then were worms. The last scientific publications Bush produced were comparisons of the worm specimens from Alaska with those found near Bermuda: "Description of New Species of Turbonilla of the Western Atlantic" in *Proceedings of the National Academy of Sciences*, Philadelphia, 1907, "Turbicolous Annelids . . . from the Pacific Ocean," 1905, and "Descriptions of New Serpulids from Bermuda," 1910. These works too are the standard reference materials describing the animals.

Bush suffered a mental breakdown some time after that. By 1911, she was no longer working in research but in arranging materials for museum exhibits, and in 1914, she entered the Hartford (Connecticut) Retreat, a home and hospital for mental patients. She then lived for a time (1920 to 1924) in Farmington, Massachusetts, returning to the Hartford Retreat in the 1930s. She died of pneumonia on January 19, 1937, while living there.

Starting as one of the many female employees in the U.S. Commission on Fish and Fisheries, Bush was one of the two 19th-century women who managed to achieve research careers in a field where women were distinct outsiders. MARY RATHBUN, who continued her career at the National Museum of Natural History, was the other. The late 19th century was not a good time for women who aspired to scientific careers. Those who were really ambitious and had good mentors (Verrill, in his way, was such a mentor) were the only ones who could continue in meaningful work.

C

⊠ Cardano, Girolamo
(Geronimo, Gerolamo)
(1501–1576)
Italian
Mathematician, Physician

Girolamo Cardano was internationally known for both his mathematics and his medicine. Born in Pavia on September 24, 1501, he was the illegitimate son of Fazio Cardano, a local jurist and legal scholar, and Chiara Micheri. Since he was recognized by his father, Cardano was educated in medicine and received degrees at the universities of Pavia in 1520 and Padua in 1526. Seeking a wider audience, he soon moved to Rome, where his medical reputation made him much sought after. A talent for attention-getting soon made him a popular physician; he was referred to as second only to his contemporary, Andreas Vesalius, and he thought of himself as a worthy successor to Archimedes in applied physics.

Leonardo da Vinci's friendship with Cardano's father had brought the son into various engineering projects in the 1520s for which Leonardo sought help with mathematical questions. In turn, Cardano was involved in some large constructions instrumental in bringing to light a number of ideas that would later be used in geology and evolution. Observing construction of fortifications and redirecting watercourses led

Girolamo Cardano, philosopher, doctor, and Italian mathematician *(Roger Viollet/Getty Images)*

him to conclude that mountains are formed and shaped by water erosion. The fossils found in mountains suggested that at one time that area was under the sea and had been lifted up. He attempted explanations for this and for the water cycle. He explained the cycle of rivers running downhill to the sea, evaporation of water from the sea to clouds, the formation of rain and rivers, as a simple, normal process that needed no complex system of tunnels through mountains or any other exotic idea of the sort that had been proposed over the centuries to explain the source of terrestrial water.

Pursuing an academic career in mathematics took Cardano back to Pavia where he served as a professor of mathematics from 1543 to 1560, and then to Bologna for the next six years. Charged with heresy, he lost that post, and after recantation, moved to Rome in 1571 to live on the stipend granted by Pius V, until his death on September 21, 1576.

Two encyclopedias of natural science are credited to him, *De subtilitati libri* XXI (1550) and *De rerum variete* (1557). These works are collections of observations (some of them fabulous) and contain a bit of everything, including some notes that might be unpublished works of Leonardo. His other publications include *Ars magna* (Great skill) (1545) and *Liber de ludo aleae* (Book on games of chance). Like Leonardo da Vinci, much of Cardano's work was known to others through his letters. They were eventually published long after he finished them. *Ars magna* and *Liber de ludo aleae* were finally printed in his collected works, *Opera omnia* (Complete works) in Leiden in 1663.

William Carpenter *(Courtesy United States Navy)*

⊠ Carpenter, William Benjamin
(1813–1885)
English
Doctor, Natural Scientist

William Carpenter is best known in connection with the voyage of the HMS *Challenger*. Born in Bristol on October 29, 1813, he was the son of Lant Carpenter, a Unitarian minister and schoolmaster who apprenticed his young son to a doctor. While a medical student, William accompanied a patient on a trip to the West Indies in 1830 and on his return to England enrolled in the University College of London. After graduating, he joined the Royal College of Surgeons in 1835 and then worked briefly in Edinburgh, Scotland, as a physician. Moving back to the south, he married Louisa Powell in 1840 and worked as a doctor in Bristol before moving his family to London in 1844.

By 1845, he was a Fullerian Professor of Physiology at the Royal Institution, and professor of forensic medicine at University College. Elected a fellow of the Royal Society in 1844, Carpenter was its medallist in 1864.

Carpenter's extensive writings on physiology date from his earliest work as a physician in 1839. By 1854, his *Principles of General and Comparative Physiology* was in its fourth edition and a standard in the field. It was followed by *Principles of Mental Physiology* in 1874. In that work, he introduced new ideas on how the nervous system worked and what constituted thought. Carpenter wrote a major review of Darwin's *Origin of Species* in 1857. In it, he accepted the basic principles of evolution and Darwin's exposition of them, but maintained the separateness of humans and believed in man's unique creation.

An excellent teacher and mentor, he was always encouraging others in the study of marine science. Carpenter's 1862 publication *Introduction to the Study of the Foraminifera* was the standard work on the subject, one that is essential in petroleum geology. Carpenter's marine studies included dredging operations near Ireland and Scotland during the summers from 1868 to 1871, when his associate was CHARLES WYVILLE THOMSON. The published work included a major exposition of the Crinoidea (sea stars).

The organization of the *Challenger* expedition occupied him for years. The voyage of the HMS *Challenger* (1872–76) was the culmination of the efforts of many people, and Carpenter was among the foremost scientists promoting this expedition. The aim was an extensive trip to study the world's oceans in every discipline available to the science of the time. Notable among those who sailed on this voyage were JOHN MURRAY, WILLIAM DITTMER, and CHARLES WYVILLE THOMSON. William Carpenter did not sail on it, but his son Philip Herbert Carpenter continued his father's work on crinoids and included finds brought back by the *Challenger* expedition. Philip Carpenter's specialty was the *Comatulae*, feather stars.

Carpenter was a founding member of the Marine Biological Association. His interest in the deep sea led to studies of oceanic physics and deep ocean circulation. His great body of pub-

lished work was a directing force in the development of 19th-century science. Carpenter died on November 19, 1885, in London.

⊠ **Carson, Rachel Louise**
(1907–1964)
American
Biologist, Ecologist

Carson's work was instrumental in the creation of environmental awareness. She was born on May 27, 1907, in rural Pennsylvania and graduated from the Pennsylvania College for Women (now Chatham College) in 1929; later, she studied at Woods Hole Marine Biological Laboratory and then at Johns Hopkins University for an M.A. in zoology in 1932.

Finding a permanent position in science was difficult during the economic depression of the 1930s. She wrote articles on natural history for a Baltimore newspaper part-time, and by 1936, Carson was editor-in-chief for the publications of the U.S. Fish and Wildlife Service. Aiming at the general public as audience, she inaugurated a career as a naturalist with an essay in the *Atlantic Monthly* in 1937, entitled "Undersea." This article was followed by *Under the Sea Wind* in 1941 and *The Sea Around Us*, a National Book Award prize winner in 1952. She was supported in her writing by CHARLES WILLIAM BEEBE. Will Beebe was then the curator of birds at the Bronx Zoo, now the Wildlife Conservation Center in New York. He was a nationally recognized science writer and conservationist. The prize-winning book made her famous and enabled her to leave the government service and concentrate on writing for the general public.

Carson's basic view was that humans, though part of the environment, had great potential for changing it, possibly irreversibly. Her book, *Silent Spring* (1962), warned against the misuse and excessive use of pesticides, noting their long-term effects on the biosphere. She was certainly viewed

Rachel Carson *(Beinecke Rare Book & Manuscript Library, Yale University)*

as an enemy by the chemical industry and as an alarmist by others. Carson was gaining international recognition for her work when she died of breast cancer April 14, 1964.

Chappe d'Auteroche, Jean-Baptiste
(1728–1769)
French
Astronomer

A capable mathematician and staff member of the Paris Observatory, Chappe was a native Parisian born on March 23, 1728. He was an alumnus of the College Louis-le-Grand, a noted institution for the education of future scientists. His background and his mathematical ability made him a natural candidate on the astronomy team that the government of France sent to Tahiti to observe the transit of Venus moving across the disk of the Sun. Viewing such a relatively uncommon astronomical event was a means of checking the accuracy of other astronomical sightings and measurements. It was also an opportunity to field-test new equipment. Every seagoing nation wanted to try navigational instruments and make its own measurements to be sure that they were correct. This was more than just national pride; they didn't trust each other. The expressed purpose of this voyage was the determination of longitude. Chappe worked on the determination of latitude using meridian altitudes of specific stars, and of longitude using lunar eclipses and oscillation of stars as fixed points. The principal piece of new equipment used was a chronometer built by Ferdinand Berthoud.

After his return from the Pacific, Antoine Lavoisier suggested a series of chemical experiments to be conducted on seawater in various locales. Chappe was again chosen to be the scientist on a voyage, this time to the coast of Baja California. He was to be the chemist on board and to continue his longitude calculations based on the readings taken from the western Pacific Ocean. The voyage was a disaster, and he was lost in a shipwreck.

Chappe's work was never published independently, but was appended to Ferdinand Berthoud's work, "Traite des horloges marines" (Treatise on marine clocks) that was published in 1773. Berthoud built the chronometer that Chappe tested on his Pacific voyage. This seagoing clock worked well and was used for the excellent prediction of lunar and solar events. It was one of the many useful additions to the growing number of tools available for marine exploration and accurate navigation.

⊠ Charcot, Jean-Baptiste
(1867–1936)
French
Explorer

Jean-Baptiste-Étienne-Auguste Charcot took part in the exploration of the polar regions and, like ALBERT I of Monaco, whom he admired and emulated, was an enthusiast and source of funding for scientists. The son of the famous neurologist Jean-Martin, Charcot was born in Paris on July 15, 1867. He trained in L'Hôpital de Paris and followed his father into medicine, specializing in neurology. His writings in that field date from 1887 to 1901.

Charcot always loved the sea, and by the time he was 35, he tired of "being his father's son," and devoted himself to marine science. In explanation for his meticulous attention to work in oceanography and the methodology he brought to his new field, he said, "I am a marine scientist."

His first exploratory sail in the Atlantic was to the latitude of the Faeroes in 1901, and then, in the following year, to Jan Mayen Island. By 1903, he was ready for polar waters. He commissioned the construction of the research vessels the *Français* and the *Pourquoi Pas?* These were the first research ships built in France; all others had been built for other purposes and refitted. The *Français* was built with reinforced bow and hull to withstand ice pressure. In spring of 1903, the cutter sailed to the Antarctic on the first French expedition to that region. The immediate aim was to search for Otto Nordenskjöld, an explorer who had been reported missing. Before the expedition, Charcot had exhausted his funds and turned to the government to help finance the project. While he did receive support from the Académie des sciences, it was the publicity given to him by the newspaper *Le Matin* that raised money for the expedition.

The ship that Charcot directed was the floating laboratory of the École pratique des hautes études (Practical School of Advanced Studies) and was often used by the Musée national d'histoire naturelle (National Museum of Natural History). The scientific crew included physicists, oceanographers, biologists, and geologists. The expedition returned in 1905 after making extensive maps of the region and collecting and analyzing samples of water and organic material. Since their original vessel had been damaged and then sold in Argentina, they returned by transport ship. Nordenskjöld was found by a group of Argentineans.

The second French Antarctic voyage was carried out in the *Pourquoi Pas?* and lasted from 1908 to 1910. Charcot then led several surveys in the Atlantic and the Arctic. In 1932, he and his ship were part of the "Polar Year" expedition to Greenland. The Polar Year was a period in which several nations with polar expertise all launched expeditions to observe and experiment in the Arctic. This was a beginning in international cooperative scientific efforts.

Charcot returned to Greenland to find Paul-Émile Victor and his expedition. Victor and his two companions were ethnographers who had wintered in Greenland. They were unreachable, and several explorers assumed that they were stranded and starving. The *Pourquoi Pas?* was lost in a storm on September 18, 1936. One man survived to tell the story, and Charcot's death was national news. Victor and his companions had survived the winter near Godthaab, Greenland's capitol.

The legacy of this scientist-explorer and his group influenced many fields in ocean science— bacteriology, botany, geology, glaciology, magnetics, mineralogy, tides, and zoology. They created two new subspecialties, submerged geology and geological oceanography. The major publications resulting from their efforts were the books *Expédition antarctique française 1903–05* (The French Antarctic expedition of 1903–05); A. *Sciences naturelles, documents scientifiques* (Natural science, scientific documents); and B. *Hydrography et physique du globe* (Hydrography and physics of the globe). These were published serially in 17 sections. Another book detailing the second voyage was *Deuxième expédition antarctique française* (Second French Antarctic expedition), published in 25 sections.

⊠   **Cleve, Per Teodor**
(1840–1905)
Swedish
Chemist, Botanist

Born in Stockholm, Sweden, on February 10, 1840, Cleve was educated at the university in Uppsala, where he received his B.A. in 1858 and a

Ph.D. in 1863, both degrees in chemistry. He then accepted a professorship there. His major area of study was mineralogy, but botany had always been his serious hobby. Because of his interest and outgoing personality, he became Sweden's leading natural scientist and was mentor to a large group. As part of his public service to the scientific community, he served as president of Sweden's Nobel Prize committee from 1900 to 1905. His work in mineralogy and chemistry led to the discovery of holmium and thulium in 1879. He continued the work of Karl Mossander, the discoverer of several rare earth metals, and with careful chemical analysis Cleve proved the existence of scandium, a new element predicted by Dimitri Mendeleev and called *ekabor* some 18 years earlier. Mendeleev arranged the then-known elements into a comprehensive chemical chart that included gaps for those elements still undiscovered.

Cleve devoted his last 15 years to botany. Starting his studies with freshwater algae, he soon became an expert on diatoms. Applying careful methodology to the study of these marine diatoms, he arranged them in a systematic chronological sequence. This sequence is used to demonstrate movements of ocean currents both present and ancient. It is also essential information used in petroleum geology because certain diatom deposits characterize oil-bearing rock. Cleve continued to work on his diatoms up to his death in Uppsala on June 18, 1905.

⊠   **Conklin, Edwin Grant**
(1863–1952)
American
Biologist, Embryologist

Edwin Conklin, born on November 24, 1863, the son of a country doctor, is remembered for his pioneering work in embryology. Originally from Waldo, Ohio, he was educated locally and then received a B.A. degree in biology from Ohio Wesleyan University in 1886. He continued at Johns

Hopkins University, where he studied for the "new degree," a Ph.D. The new biology chairman was W. K. BROOKES, and Conklin received his degree in 1891. The organism he studied for his thesis work was *Crepidula*, a limpet or marine snail. He tracked the development of its embryos and identified the regions of embryonic cells that would display patterns pointing to evolutionary relationships to other organisms. In the summer of that year he was at the marine research center at Woods Hole, Massachusetts, as was E. B. WILSON, who had done similar work on *Nereis*—an annelid. On comparing their work, they found similarities in the two animals that had been previously thought only distantly related. In 1908, the president of Princeton, Woodrow Wilson, invited Conklin to be head of the biology department. He held that position until his retirement in 1933.

The mosaic theory of development was an integral part of Conklin's research. He believed that specific areas of the fertilized egg always lead to one or another specific structure in the developing embryo. A firm believer in experimental biology, he stressed this as the essential part of embryology and continued that work throughout his life. Conklin wrote numerous papers both for the scientific and popular literature that explained and defended CHARLES DARWIN's theory of evolution. These include *Heredity and Environment in the Development of Man* (1915), *The Direction of Human Evolution* (1922), and *Man, Real and Ideal* (1943). He died in Princeton, New Jersey, on November 20, 1952.

Conklin maintained an association with the Marine Biological Laboratory in Woods Hole throughout his life. The continuing work on cell lineages connected evolutionary biology to its offshoot, embryology. This much-honored man belonged to numerous scientific societies and was president of the American Society of Zoologists, the American Academy of Arts and Sciences, the American Philosophical Association, and the National Academy of Science.

⊠ **Cook, James**
(1728–1779)
English
Navigator, Explorer, Mapmaker

James Cook was born to James Cook Sr., a ploughman, and Grace Pace, a farmer's daughter on October 27, 1728, in Marston, a village near Whitby, England. After an elementary education in local schools, the future navigator and self-made scientist was apprenticed at age 17 to John Walker, a Whitby shipowner and sailing master. The Walker house is now a Cook Museum. Walker's apt pupil was soon sailing on the regular coal run from the northeast coast of England to London and back. In 1755, he was offered command of a ship but joined the Royal Navy instead. Rising in rank, he was attached to the American station when the Royal Navy was surveying the St. Lawrence River and Gulf. There, Samuel Holland taught him the mathematics necessary for surveying and navigation. Cook married Elizabeth Shadwell in 1762, and the next year, once again in North America, he was involved in the surveys of Newfoundland.

The charts made on this voyage were favorably received by the Admiralty as was his account of a solar eclipse by the Royal Society of London. On the basis of these notices, he was chosen to command an expedition to Tahiti to observe the predicted 1769 transit of Venus (the crossing of the planet between Earth and Sun). It was understood that the secret objective of the voyage was to observe everything that might be of interest while in Polynesia.

The ship HMS *Endeavor* sailed in July 1768 and arrived in Tahiti in April 1769, in time for the transit on June 3. Cook then went on to chart the island groups and carry out the secret mission, to look for a continent south of latitude 40° south (Terra Australis). This rumored continent south of India interested the Admiralty. If it was there, it had to be proven and claimed for England. The captain was also to chart the coast of New

James Cook, among other accomplishments, charted the coast of New Zealand, claimed the east coast of Australia for Britain, surveyed the coast of Antarctica, and extensively surveyed the Pacific and Southern Oceans and their islands. *(Hulton/Archive/Getty Images)*

and his entourage of eight people and two dogs inadequate. The Admiralty refused to remodel the ships to increase space between decks because it would raise the ship's center of gravity and destabilize it. Other naturalists were appointed, REINHOLD and GEORG FORSTER, a father-and-son team. This voyage took the party to 71° 10′ South Latitude, where they again found no massive continent. They did accurately chart several Pacific island groups, Tierra del Fuego, and the South Sandwich Islands. This voyage is significant because Cook, using the data he and Banks had collected on the earlier voyage, lost no one to scurvy. Banks advocated citrus; Cook opted for the cheaper sauerkraut, but the effect was the same. The crew did not suffer from the vitamin deficiency problem endemic on long voyages. In addition to collection of animal samples, this trip established the efficacy of accurate clocks as a means of determining longitude.

The Royal Society was extremely pleased with the results of the longitude experiments and elected Cook a fellow in 1776. In July of that year he left on the third voyage, sailing in the *Resolution* and the *Discovery*. The mission was to look for the Northwest Passage, the rumored path through the Americas to the Indies on its western, or Pacific side. The ships sailed east, traversed the Pacific, and discovered the Hawaiian archipelago. This island group was named Sandwich Islands in honor of the First Lord of the Admiralty. In March 1778, Cook sailed to the northwest coast of North America and found no outlet for the passage. Since winter ice made further sailing difficult, the ships returned to Hawaii for the winter. There, on the west coast of the island of Hawaii, this extremely able commander, who had had a record of good relations with indigenous people in Australia and New Zealand, was killed on February 14, 1779 by the Hawaiians in an argument about a stolen longboat. His nation mourned him.

Cook's writings consist of one original work, *A Voyage Towards the South Pole and Round the*

Zealand, known since Abel Tasman, sailing for the Dutch, had reported its existence in 1642.

Cook, with Joseph Banks and Daniel Solander, the naturalists on board, did not find Terra Australis but did chart New Zealand and the eastern coast of Australia and collect a vast store of geographical, botanical, and ethnological specimens. They returned in July 1771. The success of this voyage set the scene for another, again to find the imagined continent. This time Cook used two renamed Whitby "cats," the *Resolution* and the *Adventure*. They were typical wide-bottomed heavy cargo carriers used to transport coal, the type of ship on which he learned his craft.

Banks declined the offer to be the onboard naturalist. He found the space allotted to him

World (1777). The Hakluyt Society published his ships' logs in three volumes between 1955 and 1967 under the title *Journals of Captain James Cook on His Voyages of Discovery*.

⊗ Cousteau, Jacques
(1910–1997)
French
Explorer, Environmentalist

Jacques-Yves Cousteau was born on June 11, 1910, near Bordeaux, France, and educated at the French Naval Academy. His original intent was to be a naval aviator. That career goal ended with a car crash that nearly killed him. As part of his rehabilitation, he began to swim vigorously to strengthen his damaged arms and "fell in love with the sea." By 1943, he and Émile Gagnan—an engineer whose association with Cousteau continued for years after the war—had designed the "aqualung." This was based on the compressed-air cylinder that had been invented in 1933. Using this cylinder directly almost killed Cousteau. Gagnan devised a better breathing mechanism that allowed a diver to move freely underwater without air lines to the surface. They called it SCUBA, the acronym for "self-contained underwater breathing apparatus." Cousteau's first dive with the new machine was to a depth of 18 m (60 feet). After the war, he was named Chevalier de la Légion d'Honneur (Legion of Honor) for his daring work in early filming and the use of SCUBA to locate and defuse German mines.

In 1950, Cousteau bought a former mine sweeper and renamed it *Calypso*, after the nymph of Greek mythology. The vessel was outfitted as a floating laboratory that included television equipment. His videos of the bottom of the Red Sea, made in 1952–53, were the first color photographs of a seabed. Cousteau's filmed documentary of the find of an ancient Greek ship near Marseilles was spectacular. The remains of this freighter and its wine cargo are on display in Marseilles in a museum near Vieux Port (the ancient harbor). The *Calypso* was his base for many other documentaries.

Cousteau wrote a number of popular books about marine subjects; two favorites are *The Living Sea* and *Amazon Journey*. His documentaries were presented on public television, and two full-length films, *The Silent World* (1956) and *World Without Sun* (1966), won Academy Awards. Other projects included the Cousteau Society, an environmental group based in the United States, and the Conshelf Project, an experiment in long-term underwater living. Triggered by the fact that divers can work only for very short periods at depths of more than 90 m (300 ft), the habitat was placed at that depth. The project studied the rate of habituation to such depths and the

Jacques Cousteau (right) with two colleagues
(SIO Archives/UCSD)

ability to work in that environment. Unhappily, it was put aside after the plane crash in 1979 that killed Cousteau's son Philippe, a major figure in Conshelf.

Appointed director of the Océanographic Museum in Monaco, Cousteau held that post from 1957 to 1988, when he resigned to give himself more time for environmental projects. He was a member of the Académie française, Académie des sciences, and the National Academy of Science in the United States, among others. He received many medals and honorary doctorates and awards. Critics emphasized his lack of scientific training. However, he continued to travel, make films, replacing the first *Calypso* with another after the sinking of the first, and pleading for clean, safe, environmentally sound use of the sea. Cousteau died in Paris on June 6, 1997.

⊠   **Cushman, Joseph**
(1881–1949)
American
Micropaleontologist

Joseph Cushman, son of Darius and Jane Pratt Cushman, was born on January 31, 1881, and grew up in rural Bridgewater, Massachusetts. At his local college, Harvard, his early interest in botany was eclipsed by paleontology. He received a B.S. degree in biology in 1903 and then went to the Museum of the Boston Society of Natural History as its curator.

Cushman spent the summers of 1904 and 1905 at the U.S. Fish Commission Laboratory in Woods Hole, Massachusetts, where MARY RATH-BUN persuaded him to study the Formanifera specimens collected by ALEXANDER AGASSIZ on the *Albatross,* which had received no systematic attention up to that time. Cushman identified 10 families of these curiosities dating from the Cambrian to the Holocene, believing that he had seen almost all of the foraminifers in the world, but a wealth of new specimens continued to be found. He spent years on this work, finally producing a first publication in outline in 1927. This skeletal form eventually became a book, *Foraminifera; Their Classification and Economic Uses.* By then, there were 45 families of these organisms. The work was a classic text, periodically added to and eventually reissued in four editions.

In 1912, Cushman left the museum in Boston for the U.S. Geological Survey. This gave him a better base for his growing expertise on foraminifers. By 1918, he was using his "curiosities" as indicator species in the identification of geological strata and periods. This research was extended to the use of planktonic species as a means of determining geological stratigraphic layers. The fossil plankton were used to typify and date the layers of sedimentary rock in which they were found.

The work on microfossils is of enduring importance in paleoecology and stratigraphy. Its practical application is the identification of appropriate geological structures that would indicate the possible presence of petroleum.

⊠   **Cuvier, Georges, Baron**
(1769–1832)
French
Anatomist, Zoologist, Paleontologist, Historian of Science

Cuvier's father was a Swiss mercenary serving the French government in Montbéliard, which was in the duchy of Württemberg when Georges-Léopold-Chrétien-Frédéric-Dagobert was born on August 23, 1769. The home of this Protestant family was in a part of France whose government changed periodically according to the political climate.

An outstanding student, Cuvier was sent by the duke of Württemberg to the Carolinium, the duke's newly founded university in nearby Stuttgart. Upon fulfilling the requirements for a degree in 1788, he expected but did not receive a position in the ducal government. Instead, he became a tutor in a noble household in Normandy for six years. This position gave him access to the libraries and botanical gardens of the region. Summers spent at the beach in Fécamp gave Cuvier access to the birds and littoral (coastal) animals. He filled notebooks with drawings and very detailed descriptions of the creatures he found. Although he avoided classification schemes and theories of relationships between animals, by 1793 he was writing of "descent by degrees."

When Montbéliard was annexed by republican France in 1793, Cuvier became a naturalized French citizen and immediately wrote to ÉTIENNE GEOFFROY SAINT-HILAIRE, professor at the Musée d'histoire naturelle (Museum of Natural History). Geoffroy invited Cuvier to come to Paris. The first scientific lecture Cuvier presented there was about the invertebrates of Normandy.

While claiming to be uninvolved in theories of descent, what we would now call taxonomy, Cuvier designed a map of relationships of organisms that was used by JEAN-BAPTISTE LAMARCK by 1796. The lack of trained zoologists meant that those with experience could rise rapidly, and that is just what Cuvier did. He received professorship at les Écoles centrales (the revolutionary replacement for the universities) and the assistant professorship of animal anatomy at Musée d'histoire naturelle. Part of his salary included the use of a house in the Jardin des plantes (the Botanical Garden). Because it was Cuvier's house, it became the center for scientific activity for all France and the rest of Europe. The referrals Cuvier made were essential first steps for many aspiring young scientists. The musée and Cuvier were the center for the administration of the scientific and research centers in all parts of France. The training and instruction of all French scientists was centralized and directed by this institution and the man associated with it. The extensive library that Cuvier had accumulated, reputed to contain more than 19,000 volumes, was open to any student of nat-

Georges Cuvier *(E. F. Smith Collection, Rare Book & Manuscript Library, University of Pennsylvania)*

ural science. Many took advantage of this incredible collection.

Cuvier's publications include three general works on zoology and one massive one on fish. The first volume of his *L'Histoire des poissons* (History of the fishes) appeared in 1828, the ninth, shortly before his death on May 13, 1832, and the 22nd, which was edited by a colleague, in 1849. This work is the basis of modern ichthyology. It was intended to be part of one massive compendium of comparative anatomy. This aim was never realized as it would have taken several lifetimes to accomplish. "Historian of science" is another title Cuvier assumed for himself. He wrote biographies of scientists, starting with the ancient Greeks through most of European history.

In terms of scientific theory of his day, Cuvier was a firm believer in the catastrophist view of living and fossil organisms. Catastrophists believed that the extinction of organisms, as evidenced by the fossil record, was accomplished by Noah's flood or an event analogous to it. A catastrophe such as a massive flood or a volcanic eruption was the event that created new species. This view was both politically and religiously important. Since the Bible was taken to be factual history, the extinction of fossil forms had to be the result of Noah's flood. To believe otherwise was contrary to the established religious law and therefore contrary to secular law of whatever country upheld this as truth. Cuvier, as a Protestant in a Catholic country, always tried to position himself so that he did not offend his secular superiors. It is therefore ironic that his paleontological research provided the key examples for the Darwinian theory presented in 1857. CHARLES DARWIN, in his work, set aside religious doctrine and believed that evolution was the result of chance changes in genetic structure that favored survival. As Cuvier attempted to support the view of separate creation for each species, he turned against Lamarck and

Geoffroy St.-Hilaire in 1804. This was to some extent a refutation of his own work *Hydrologie* (1802), in which he describes fossil invertebrates and places them in a pattern that illustrates evolutionary change. On paper, however, Cuvier held strictly to the biblical chronology. He continued his enmity to Geoffroy St.-Hilaire until he died. When Lamarck died, Cuvier's funeral tribute so belittled Lamarck's work that it took more than a century for scientists to reexamine his pioneering efforts and give him due credit for them.

Administration of the large network of rapidly growing departments of natural science both in Paris and the provinces greatly occupied Cuvier. He was politically flexible and was a councilor of state for both Napoléon and Louis-Phillipe. In 1831 he was raised to the peerage, a rare honor for a Protestant. Unfortunately, he was not a happy man. His wife and all his children predeceased him, and he was constantly involved in the minutiae of administration. He died in the house in the Jardin des plantes that was his home, library, and laboratory on May 13, 1832.

D

⊠ Dana, James Dwight
(1813–1895)
American
Geologist

An internationally known scientist, James Dana was the foremost American geologist of his day. He was born in Utica, New York, on February 12, 1813, to James Dana, a saddlemaker, and Harriet Dwight Dana. Dana was educated locally and then sent to Yale University. At Yale, he met the chemistry professor, Benjamin Silliman, who encouraged him in his geologic studies. After graduation, Dana spent 1833 and 1834 on the USS *Delaware* as the teacher for the midshipmen. His first scientific publication was the description of the 1834 eruption of Vesuvius, seen from the *Delaware*. After his voyage, he returned to Yale and Silliman's laboratory. The combination of chemistry and geology resulted in Dana's definitive work, *System of Mineralogy*, which appeared in 1837.

On the recommendation of the botanist Asa Gray, Dana was appointed as one of the scientists on the Charles Wilkes expedition of 1838–42. That American venture sought to circumnavigate the globe, explore the natural history of Polynesia, and confirm the existence and map the contours of a possible Antarctic continent. After Joseph Couthouy left the expedition in Sydney, Australia, Dana served as both marine

James Dana *(E. F. Smith Collection, Rare Book & Manuscript Library, University of Pennsylvania)*

zoologist and mineralogist on the voyage. This expedition served for Dana as the *Beagle*'s voyage served Darwin; it was the data-gathering time for his life's work.

Dana's atlas and reports of the voyage were published as separate volumes; *Zoophytes* (1846), *Geology* (1849), and *Crustacea* (1852–55). The work on the organisms that Dana and his contemporaries called zoophytes (now known as coelenterates) was new to the science of his day. The coral plants and animals had just been discovered. Dana's classification of coral-making organisms became the authoritative work on the subject and his thoughts on corals were published in a separate volume, *Corals and Coral Islands,* in 1872, based on his first observation of a coral island in 1839. It was in 1839 that he found a newspaper in Sydney that had printed CHARLES DARWIN's explanation of atoll formation, fringing reef, and barrier reef structure. The paper had reprinted the text of a lecture Darwin presented in London shortly after returning from his round-the-world voyage. According to Dana's notes, "it threw a flood of light over the subject." Dana had grasped the correlations between the animals and the structures they built with the geologic changes in the shape of Earth's crust as a result of volcanic activity.

Described as a uniformitarian in geology and a catastrophist in biology, Dana attempted to retain a biblical worldview. Uniformitarians were a group of geologists who believed that all the processes that formed the Earth took millions of years. The layering of sedimentary rocks was the classic example. The catastrophists believed that every geologic event was a sudden and violent one, such as a volcanic explosion or lava flow. Dana believed that volcanic activity was the driving force for change in the Earth's crust and that these violent events created changes in surrounding flora and fauna. He had visited many volcanic regions after the Wilkes voyage and had outlined reasonable steps to explain the volcanic process both on land and under the seas. It took him some time to incorporate Darwin's thinking about change in life forms. As a part of the generation that believed that it had to choose between science and traditional religion, this was a gradual and sometimes a troubling process. He

worked to maintain the traditional explanation for differing species as a series of separate creations, each following a catastrophic event.

After retiring from Yale in 1890, Dana continued to edit the *American Journal of Science* that he and Silliman had founded. He died in New Haven, Connecticut, on April 14, 1895.

⊠ **Darwin, Charles Robert**
(1809–1882)
English
Biologist, Geologist, Natural Scientist

Charles Darwin was born on February 12, 1809, at the Mount in Shrewsbury in western England; his was a famous family. Erasmus Darwin, a physician and natural scientist who wrote about the inheritance of characteristics, was his grandfather. Charles was tutored by his sisters after his mother, Susannah Wedgwood, died when he was eight. He was later sent to a day school in Shrewsbury, where he was an indifferent student. Robert Darwin, his father, who was also a doctor, then sent him to the University of Edinburgh in 1825 to train in medicine. The subject revolted him and he avoided the medical lectures, but he did attend those given by Robert Grant, a biologist who followed the theories of JEAN-BAPTISTE LAMARCK, that animals changed, and that these changes were inheritable. Robert Jameson, with whom he collected marine organisms in the Firth of Forth in Scotland, was another influential Edinburgh acquaintance. When Darwin returned to Shrewsbury in 1827, his father tried again to find Charles a career. This time he was sent to Cambridge to prepare for the clergy; that proved to be another unsuccessful choice.

One happy outcome of Darwin's stay at Cambridge was his introduction to the work of the geologist Adam Sedgwick and his meeting with John Stevens Henslow, who inspired him with a passion for natural science. After receiving his ecclesiastical degree in 1831, and while waiting for appointment to a country parsonage, he was invited—thanks to Henslow's intervention—to join the admiralty surveying ship the HMS *Beagle* as an unpaid naturalist and gentleman companion to the captain, Robert Fitzroy. They sailed on December 27, 1831, to survey and chart the coast of South America, and various Pacific islands. After a five-year voyage, though Fitzroy and his naturalist companion were barely speaking, Darwin returned as a man of science who would change geology and biology forever.

Darwin moved to London, where his brother Erasmus introduced him to the scientific social life there. He married his cousin Emma Wedgwood in 1839, and with his family—there were eventually 10 children—moved to Downe in Kent. Once established there, he continued his research, rarely leaving the comfortable house. Much has been said of Darwin's "retreat" from the scientific community in London. He moved largely because he was increasingly ill in London, and his country place improved his comfort. He had been variously diagnosed with malaise, hypochondria, depression, or all three. The likeliest cause of his lassitude and intestinal problems was Chagas' disease; he was certainly exposed to it in South America, and he did die of heart problems, another long-term complication. His doctors would have had no way of recognizing it or treating it then. Seven of his children lived to maturity, and his four sons pursued careers in science or technology.

Darwin's work was pivotal in geology, biology, and evolution. However, he first thought of himself as a geologist. Geology was a rapidly changing field when he first encountered it. In the early 1800s, the explanation for the stratification of Earth, and the existence of fossils of extinct organisms in Earth's layers, was that a succession of catastrophes—destructive events— were each followed by new creations. The English expounders of this version of Earth history were William Buckland, William David Conybear, and Adam Sedgwick. CHARLES LYELL,

Charles Darwin *(E. F. Smith Collection, Rare Book & Manuscript Library, University of Pennsylvania)*

whose first volume of *Principles of Geology* (1830) challenged this theory, was taken by Darwin aboard the *Beagle*. Lyell explained stratification, erosion by wind and water, and elevation or subsidence of landforms as continuing processes, including catastrophic events. Both were part of geologic processes. Darwin brought the first volume of Lyell's work aboard the *Beagle*. His first use of Lyell's geology was in the Cape Verde Islands. There he examined the rocks of a mountain whose base was volcanic in origin. This base was covered by a limestone layer, with the fossils of marine animals embedded in it. Another volcanic layer covered the limestone. Darwin reasoned that if Lyell was correct, a volcanic cone subsided into a shallow sea, acquired over a period of time the characteristic deposits of marine debris, and then was relatively quickly raised above the water level and later experienced another volcanic eruption. It was a better explanation of the evidence than was a series of new creations. St. Paul's Rocks in the mid-Atlantic were other anomalies that catastrophism alone could not explain. They are neither coral nor volcanic. They are now believed to be exposed bits of Earth's mantle.

South America offered other geological evidence. Darwin noted the similarities between lava and granite. Darwin looked at rocks after reading Lyell and saw that some of them were formed by sedimentation; others were the result of volcanic eruption or of the intense heat that accompanied such an event. The large crystals found in the structure of these rocks showed this. If one strikes such a volcanic rock, it foliates or cleaves, or both; that is, pieces flake off, or it breaks along a crystal face. These breaks are parallel to the axes along which elevation has occurred. This is not dependent on stratum deposition, and therefore the uniform building up of layers of earth could not be the only explanation for the forms he saw. The inevitable explanation of the formation of metamorphic rock by deformation due to heat and pressure, and the difference between those processes and the slow laying down of layers of sediment, are among Darwin's major contributions to geology.

An earthquake occurred on the western coast of South America when the *Beagle* was in the area. Darwin correlated the land rise in the area, the volcanic activity nearby, and the appearance of a new volcanic island offshore. He observed seashells—old ones, but not fossils—similar to those on the coast in beds at elevations of 400 m (1,300 ft). These finds led him to believe that the shell-bearing heights had been only (relatively) recently uplifted. The shell-bearing rocks were decreasingly evident with increased elevation, and the naturalist's explanation was that erosion had gradually removed the traces of marine organisms. Realizing that this wearing away of the rocks would have taken a long time,

he concluded that the Earth was far older than people thought it might be.

Darwin examined the effects of climate change by observing the deserted villages that had been lifted up above the snowline, but which bore evidence that they were once at elevations where agriculture was possible. Later, in the Caroline Islands, after he acquired Lyell's second volume in Sydney, Australia, Darwin saw drowned houses, giving evidence for the theory of subsidence of land areas. He drew this conclusion by connecting his seemingly unrelated observations of active volcanoes and elevation of the sea bottom and fringing reefs. Coral organisms had been described by others before Darwin saw them; they are shallow-water creatures that cannot live in temperatures much below 20°C. Thus, the presence of their remains at depths greater than about 35 m (120 ft) indicates an area that has subsided or that the sea level has risen. The Darwinian theory of the formation of coral atolls was reaffirmed in the 20th century.

Once he returned to England in 1836, Darwin wrote of his findings in the published works *Journal of Researchers into the Geology and Natural History of the Countries Visited during the Voyage of H.M.S. Beagle Round the World* (1839), *Geological Observations on South America* (1846), and *Geological Observations on the Volcanic Islands Visited During the Voyage of H.M.S. Beagle* (1844).

Established in Downe, Darwin undertook an exhaustive study of barnacles. Four volumes on the subject were published, which were the definitive study of these marine mollusks: *A Monograph of the Subclass Cirripedia*, in two volumes that appeared in 1853 and 1854; *A Monograph of the Fossil Lepadidae, or Pedunculated Cirripedes of Great Britain* (1851); and *A Monograph of the Fossil Balanidae and Virrucidae* (1854).

Pigeon breeding also occupied Darwin. This was very much a Victorian gentleman's hobby, but he approached it with the scientific question: can selection of characteristics be quantified and predicted, or is it chance? The pigeons came in a vast variety. Like any good stockman, Darwin selected specimens for desired traits and observed that they bred true. While he could not explain the mechanism (he did not know about Mendel's work on inherited traits), he knew that he could create true varieties of fancy pigeons, birds with specific desired characteristics.

His perfecting of his famous book, *On the Origin of Species by Means of Natural Selection, or the Preservation of Favoured Races in the Struggle for Life*, went on for years. Darwin was finally propelled into publishing it by Alfred Russel Wallace's paper. Wallace had come to the same conclusions as Darwin had and sent his paper on the subject to Darwin, the professional scientist's authority, for comment. At Darwin's insistence, the two works, Darwin's and Wallace's, were presented to the Linnaean Society on the same night in July 1858, and at the urging of his friends, Darwin finally permitted publication of his book on the subject in 1859.

The storm of comment *On the Origin of Species* received had been anticipated and dreaded by its author. The conclusions presented in the original edition of the book, which he explained further in later editions, were based on his observations while on the *Beagle*'s voyage. For example, he noted that the living South American armadillos were similar but *not* identical to extinct species. Again in South America, the flightless birds of the pampas (grassy plains) resembled the smaller species found farther south, but were different from the South African ostrich, which inhabits a similar grassland biome; and South American agoutis were like other South American rodents but quite different from North American rodents. He looked at oceanic organisms and found them analogous to others on the nearest continents. Thus Cape Verde animals were like African ones and Galápagos animals like South American ones. Thus, while the Cape Verde Islands and the Galápagos Islands are in similar latitudes and have similar geology, climate, and physical features, their animals resemble those of the nearest continents; Cape Verde animals are unlike those of the similarly placed Galápagos. Darwin's acute observations of Galápagos species showed that while all the islands have similar finches, they differ in size, favorite food, and bill shape, and that while all the islands have tortoises, the shell patterns differ from island to island. These and other observations led Darwin to surmise that species do change with time, and these changes result in adaptations or differences. In isolated environments, those differences become more obvious.

The first publication of these thoughts began in 1837 in *Notebook on Transmutation of Species*, and Darwin became increasingly drawn to the concept after studying comparative anatomy; thus a horse's leg, a bat's wing, a seal's flipper, and a human arm are similar in structure and function similarly. Darwin noted the similarity of embryos and the groupings of organisms into distinct forms, which led him to the concept of a single origin for all.

"Darwinism" was unacceptable to the religious establishment. Its creator was amused by this storm; he had remarked on one occasion, "I was once in the process of becoming a clergyman." After his death on April 19, 1882, several members of Parliament petitioned the dean of Westminster Abbey to allow his burial in the abbey. The honor was granted.

⊠ Delage, Yves
(1854–1920)
French
Embryologist, Zoologist

A native of Avignon, Yves Delage was born on May 13, 1854, and attended local schools, passing the *bac*, an examination that entitles one to go to university in science and literature, in 1873. He then spent two years teaching science in La Rochelle. In 1875, he moved to Paris where he studied medicine and was licensed in natural science in 1878.

Befriended by Lacase Duthiers, his mentor and teacher in Paris, in 1878 Delage was appointed director of the Sorbonne's marine biology laboratory in Roscoff—a channel fishing port. In his earliest work, his M.A. thesis, not published until 1881, in *Archives de zoologie expérimentale et générale* (Archives of experimental and general zoology), Delage described the circulatory system of crustacea. While at Roscoff, Delage completed his work for the doctorates in medicine in 1880 and in science in 1881.

Maintaining his position as director of the laboratory and professor of zoology at the Sorbonne, Delage remained in the channel region for the rest of his life. When he began his studies of sponges, many biologists still were not sure if they were indeed animals. Linking the development of the sponge egg, which is mobile, to the sessile (immobile, fixed to sand or rock) adult organism was a great step forward in the understanding of the reproductive process of this so-called simple organism. This work on the metamorphosis of sponges was pivotal. Other projects he was involved with studied the nervous system of a crustacean parasite, the development of eels and of ascidians. He was the first to artificially fertilize animal eggs, producing sea urchins that lived to maturity. Because he so strongly believed in the effect of environment, Delage has been thought of as one of the last Lamarckians: biologists who maintained JEAN-BAPTISTE LAMARCK's belief that organisms changed their structures *because* their environment was altered. Because of Delage's adherence to Lamarck's theories, he was unwilling to accept Mendelian genetics. Gregor Mendel had explained how an inherited characteristic could be passed on through several generations, reemerging in a later generation in the same form. The characteristic is not "diluted," because it is present on a specific chromosome—a gene. This special characteristic might be masked in intervening generations by a more dominant form; thus the offspring of an organism bearing a mutated gene may not display the characteristic associated with it. But if this hybrid is mated with a similar hybrid individual, the offspring of that pairing may exhibit the grandparent's characteristic difference.

After suffering a detached retina in 1912, Delage was forced to give up active research and began a new career in psychology. The statue of Delage at the Marine Laboratory in Roscoff is a reminder of his legacy to the research station, a body of work that included scientific publications, students mentored, and fictional and poetic works. Delage died at Scéaux on October 7, 1920.

⊠   **Derugin, Konstantin Mikhailovich**
(1878–1938)
Russian
Oceanographer, Zoologist

Derugin was always a St. Petersburger; he was born in St. Petersburg, Russia, on February 10, 1878, and lived in the Baltic city for the major part of his life. While still an undergraduate at the University of St. Petersburg, he took part in an expedition to the White Sea. This was just an undergraduate's cruise, but he became enchanted with the polar regions, and it determined his future interests. After graduation in 1900, Derugin was a research assistant and received an M.S. in zoology in 1909, and a lecturer's appointment. He received a doctorate in zoology and comparative anatomy in 1915, and by 1919 he was raised to the professorship. The next year he was named deputy director of the State Hydrological Institute.

Derugin's best-known early work was on the fauna of Kola Bay. The study of arctic waters and the organisms found therein occupied him throughout his career. Together with his students, Derugin outlined the zonation and biomes—environmental and physical—of the Barents Sea. Beginning in 1921, he sailed regularly on sampling missions in the Barents Sea in order to study the effects of the warm North Cape currents and the distinct zones of animal populations that these warmer waters created in the Arctic.

Derugin organized and led an expedition to the Pacific Ocean in 1923–33. The extensive voyage used fishing trawlers to sample the Sea of Okhotsk, the Sea of Japan, and the Bering and the Chukchi Seas, bringing organisms from depths of 3,500 m (11,500 ft). More than half of the specimens retrieved were unknown to the researchers.

Derugin was elected a member of the Society of Scientists of Leningrad when that body began in 1919. He had organized more than 250 expeditions and led most of those. All were in Soviet waters, and many concentrated on the arctic region. The Oceanography Institutes of Murmansk and Vladivostok were under his directorships.

In addition to his administrative duties, Derugin designed and tailored research methodology to the place being studied. The work of Derugin was significant in Russia as a step in the scientific study of arctic waters, rather than using them as endlessly exploitable resources. Outside the country, Derugin's efforts were aligned with the modern idea of communication with other researchers of his day. These scientists were truly interested in making science an international endeavor, not one based on national interests. This capable environmentalist died in Moscow on December 27, 1938.

⊠ Desor, Pierre
(1811–1882)
French
Geologist, Paleontologist

Pierre Desor was of a Huguenot family living in Germany. He was born in Frankfurt-am-Main on February 13, 1811, and educated at the universities in Geissen and then Heidelberg in the 1830s. He met and was befriended by LOUIS AGASSIZ while both were studying glaciology in Neuchâtel. They were part of the circle of young naturalists

exploring the mountains and fossil finds in Switzerland. Agassiz moved to Cambridge, Massachusetts, in 1846 and immediately suggested that Desor follow him there. Desor did so, but the move was a disaster that ended their friendship. However, Agassiz did recommend Desor for a position with the U.S. Coast Survey. Working for the Coast Survey, Desor collected and cataloged the aquatic fauna of the United States, both on the Atlantic coast and in the Great Lakes. When his brother died in 1851, he inherited the family fortune, which gave him the time to devote himself fully to the study of fossil echinoderms. His work on the subject, *Synopsis des échinides fossils*, was published in 1858.

Up to this time, Desor had been considered a Frenchman. On application, he became a naturalized Swiss citizen in 1859 and entered politics. He eventually became president of the Swiss Federation in 1874. Increasingly incapacitated by gout, he moved to a warmer climate and spent his last years on the French Riviera, studying the geology and archaeology of the region.

Pierre Desor died in his beloved Nice, France, on February 23, 1882.

⊠ Dietz, Robert Sinclair
(1914–1995)
American
Geologist

Robert Dietz was born on September 14, 1914, in Westfield, New Jersey. His father, Louis Dietz, was an engineer. He received all his degrees from the University of Illinois between 1933 and 1941; his Ph.D. was in geology with chemistry as a minor. Although the Ph.D. was granted by the University of Illinois, his doctoral research was done at the Scripps Institution of Oceanography (SIO), in La Jolla, California. On the West Coast, Dietz and a fellow student, K. O. Emery, and their mentor, Francis Shepard, first examined the submarine phosphorites found offshore.

Unable to find a university position, Dietz, who was in the army reserves, was called to active duty in World War II (1941–45) and served as a pilot. He remained in the reserves and retired as a lieutenant colonel. After the war, Dietz became founder and director of the Sea Floor Studies group at the Naval Electronics Laboratory (NEL) in San Diego, California. This led to a position as a geological oceanographer on Admiral Richard E. Byrd's last exploratory trip to Antarctica in 1955–56. The cooperative efforts of SIO and NEL led to a cruise to explore the Cape Mendocino submarine scarp (cliff) off the coast of California. Dietz and Robert Dill drew the first map of the Monterey Canyon, and the sea fan at its mouth. This undersea analog of a surface river delta shows the pattern of terrestrial sediments and how they are deposited on the seafloor.

Dietz was an early SCUBA enthusiast and, working for oil companies, made many dives to explore the recently discovered oil fields off the California coast. His appointment as adjunct professor at SIO, 1950–63, overlapped with his service at the NEL, 1954–64.

Dietz was in Japan as a Fulbright Scholar in 1953. The time in Japan was spent in part studying the transmission of sounds from a 1952 earthquake and extended to the geology of the north-west Pacific Ocean. The chain of ancient seamounts was an obvious arc leading from the Hawaiian Islands westward to Midway Island and then to Japan. By 1953, Dietz believed that something had to be carrying these ancient mountains along. He became an early advocate for the idea of continental drift and worked on the concept he named seafloor spreading. He then served with the Office of Naval Research (ONR) in London from 1954 to 1958. While in London, Dietz was introduced to AUGUSTE PICCARD by JACQUES COUSTEAU. Dietz suggested that the ONR would be interested in Piccard's submersible, the *Trieste*. This was Piccard's research vessel, a two-man diving chamber that could withstand the pressure of great depths. The

result was that the *Trieste* was used as a research tool by the ONR, and a navy lieutenant, Donald Walsh, accompanied Piccard on his record dive in the Challenger Deep, the great undersea chasm near the Mariana Islands. Dietz and Piccard wrote of this accomplishment in their book, *Seven Miles Down: The Story of the Bathyscaphe, "Trieste"* (1961).

The U.S. Coast and Geodetic Survey wished to expand its oceanographic and geologic studies to bring to marine studies the same excitement and enthusiasm (and money) that was being spent on the space program. Thus, in 1963, Dietz was invited to join this work. The research facilities of the Survey were moved to Miami and renamed the Environmental Sciences Administration, and were later incorporated into the National Oceanic and Atmospheric Administration (NOAA). Dietz's team of researchers, marine biologists, and geophysicists, a group similar to the one he assembled at NEL, worked on plate tectonics studies. They functioned as a team until 1975, when funding for geophysics and geology became increasingly difficult to find because of the national emphasis on space. Nearly all available research funding was allocated to space exploration, so much less went to Earth exploration.

After officially retiring, Dietz taught for a year at the University of Illinois, Washington State University, and Washington University (St. Louis), before accepting a permanent position as professor emeritus at Arizona State University in 1977. There he continued research and publication until his death in Tempe, Arizona, on May 19, 1995.

Dietz was a character with an eye for the dramatic. During the cold war, he managed to photograph Russian oceanographic laboratories. When in Prague, Czechoslovakia, for a conference in 1968, he took his camera into the streets as Russian tanks were rolling in and took photographs. Some of these pictures were published in the popular *Life* magazine, in September 1968. At the end of his career, he returned to an earlier area of interest, exploring meteor craters, and coined the term "astrobleme" to describe the scar on the Earth produced by meteor impact.

Dietz was awarded the Bucher Medal of the American Geophysical Union, the Gold Medal of the Department of Commerce, the Alexander von Humboldt Prize (Germany), and the Penrose Medal of the Geological Society of America. He contributed original work on the evolution of continental terraces, the origin of the continent slopes and margins, and the Hawaiian rise. His championing of the continental drift theory was important in furthering work on this concept. The book he coauthored with John Holden, *Creation/Evolution Satiricon: Creationism Bashed* (1987) was an example of his attitude toward science. It is a humorous refutation of the creationist view of Earth's history and development. Dietz published numerous works on the construction of the seafloor at crustal plate edges, and its erosion and change with earthquakes and slides. These were published in a number of important journals, such as the *Journal of Geology, Nature, American Scientist,* and *EOS: Transactions of the American Geophysical Union.* He also wrote a long autobiographical article for the amateur scientist, "Earth, Sea and Sky—The Life and Times of a Journeyman Geologist." This was published in *Annual Review of Earth and Planetary Sciences,* in 1994.

1861–69. Dittmer was then chief assistant in the chemistry laboratory at the University of Edinburgh and left after a few months for a three-year period from 1869 to 1872 teaching meteorology in Germany. He then returned to Scotland, first to Edinburgh, and then in 1873 to Anderson's College in Glasgow. Glasgow remained his professional association for the rest of his life. It was there that he conducted his researches and published *The Report on Researches into the Composition of Ocean Water* (1884). This was the definitive statement on the chemical composition of the samples brought back by the HMS *Challenger.*

Dittmer confirmed JOHANN FORCHHAMMER's principle, that the ratio of substances in the water samples remains constant, although the salinity varies. Not only did he determine exactly what each sample was composed of, but he also measured the gas absorption of which each was capable. These factors determine the limits of the species that each water environment can support. Recognized by his scientific contemporaries, Dittmer was awarded the Graham medal of the Glasgow Philosophical Society, installed as a Fellow of the Royal Society of Edinburgh (1863), and elected a Fellow of the Royal Society of London in 1882. William Dittmer died in Glasgow on February 9, 1892.

⊠ **Dittmer, William**
(1833–1892)
German/British
Chemist

Born in Ulmstadt, now in Germany, on April 15, 1833, William Dittmer was trained as a pharmacist. While working as an assistant in Robert Wilhelm Bunsen's laboratory in 1857, he met Henry Roscoe, an English chemistry student, who befriended him. When Roscoe left for a position in Manchester, England, Dittmer went with him, and they worked together at Owen's College from

⊠ **Dujardin, Félix**
(1801–1860)
French
Botanist, Microbiologist

Born into a watchmaker's family in Tours, France, on April 5, 1801, Dujardin attended local schools, where he became fascinated by Fourcroy's *Chimie* (Chemistry). Studying with his brother for the entrance exams demanded by L'École polytechnique produced an unhappy result: he failed, and his brother succeeded. Félix's second choice for a career was art; he studied with François Gérard in

Paris and was again deemed not good enough. Gérard was the official portrait painter for both the Bonaparte and the restored Bourbon courts.

An attempt at engineering brought him to a position as hydraulic engineer in Sedan. He married Clémentine Grègoire in 1823, and accepted a position in a *collège* in Tours, teaching mathematics and literature. While working as a teacher, he published on the Tertiary rock strata and its fossils found in the area. In 1826, the town increased the courses offered at the *collège*, and Dujardin began teaching chemistry and studying optics and crystallography. He also continued his drawing and published his noteworthy botanical work, *Flore complète d'Indre-et-Loire*, in 1833.

Advised by a friend to concentrate on one area of science, he chose zoology and in 1839 was finally appointed to a university chair in the geology department at the University of Toulouse. The Faculty of Sciences in Rennes invited him back in 1840 for an appointment in zoology and botany. Election to the Academie des Sciences (Paris) came shortly before he died on April 8, 1860.

Dujardin's contributions to marine science grew out of his zoological study of a new family of organisms that he called rhizopods (root feet). This work absorbed him from 1834 until the end of his life. The work on these tiny creatures grew out of a study of living foraminifera that had been called microcephalopods (tiny creatures with heads). That name was given them by GEORGES CUVIER and his students. Dujardin thought that these microscopic organisms were far less complex; that they were simply foraminifera without shells. Dujardin, a good microscopist, also disagreed with CHRISTIAN EHRENBERG that the tiny organs of microscopic organisms were analogous to large ones in more complex animals. Dujardin called the living tissue in his tiny creatures *sarcode* (flesh). He did not extend the idea to all organisms; he did not realize that what he called *sarcode*—the living material in all cells—would in a few years be known as protoplasm.

The use of microscopy in marine biology owes much to Dujardin, who designed optical equipment that greatly enhanced the use of microscopes. That, his magnificent illustrations, and the careful chemistry used in the analysis of *sarcode* fix his place in the history of science. Ninety-six of his works were published, including *Récherches sur les organismes inférieurs* (Research on lower organisms) in *Annales des sciences naturelles* (Annals of natural science). Dujardin died in Rennes on April 18, 1860.

⊠ **Dumont d'Urville, Jules-Sébastien-César**
(1790–1842)
French
Explorer, Geographer, Natural Scientist

The third son and eighth child of Gabriel Dumont d'Urville, a wealthy bourgeois judge in Normandy, was born on May 23, 1790. His mother, Jeanne, was a member of one of the oldest of French noble families. When his father died in 1797, Jules Dumont D'Urville was the only surviving son. The child was educated by his uncle, a priest, and entered the Lycée Malherbe in Caen, as a scholarship student in 1804. At 17 he joined the navy as a midshipman and in 1812 was promoted to ensign. He married Adèle Pepin of Toulon in 1815.

Early in Dumont d'Urville's naval career, he developed an interest in botany. He participated in a voyage to the eastern Mediterranean in 1820 that was in part a specimen-collecting trip, and he based his 1822 book *Enumeratio plantarum* (List of plants), written in Latin, on the specimens he brought back. Another result of this voyage was the transportation of the statue, Venus de Milo, to France. On his return, he joined the Linnaean Society, a scientific society devoted to botany, was named Chevalier de la Légion d'honneur (elected to the French military honors society), and was promoted to lieutenant.

Jules Dumont D'Urville *(Getty Images)*

On August 11, 1822, *Coquille* began its voyage of discovery with LOUIS-ISIDORE DUPERREY as commander and Dumont d'Urville as second in command in charge of the botanical and entomological research. The crew of the *Coquille* charted the coast of New Guinea, discovered several islands, and brought back a large collection of specimens. The expedition returned to Toulon after 31 months at sea.

Dumont d'Urville was promoted again, and as commander of the *Astrolabe* (formerly the *Coquille*), sailed again from Toulon on April 25, 1826. The mission was again exploration of the South Pacific, particularly New Zealand and the areas where, in 1788, Comte Jean-François de Galaup de la Pérouse, a French explorer, along with his crew and their ship, had disappeared while on an exploratory voyage.

Accompanying Dumont d'Urville in specimen collection were three onboard naturalists.

Upon their return in August 1829, Dumont d'Urville was promoted again, received by the Académie Royale des Sciences, and commanded by King Charles X to publish his account. It filled 12 volumes and five albums of drawings and charts when it was completed in 1835. Another result of this voyage was Dumont d'Urville's five-volume work on New Zealand, wherein he described in detail the life, customs, and language of the Maori people. His popular work was *Voyage autour du monde* (Voyage round the world). A third voyage began in August 1837. This was to be a study of the ethnography and languages of peoples in the area, as well as the polar regions. The *Astrolabe* was accompanied by the *Zélée*; the ships charted the coastline of New Zealand shortly after the islands' annexation by Great Britain.

Upon Dumont d'Urville's return in 1840, he attained the rank of rear admiral and the medal of the Société de géographie. Again commanded by the king, this time Louis-Philippe, to publish, he produced the first three volumes of *Voyage au pôle sud et dans l'Océanie* (Voyage to the South Pole and Oceania). He was working on the fourth volume when he was killed, along with his wife and son, in a railway accident on May 8, 1842.

Dumont d'Urville was a skillful navigator, a linguist, an efficient commander, cartographer, and botanist. His name survives in the genera of seaweeds, and among islands off the New Zealand coast and in Antarctica, where Adélie Land and a species of penguins are named for his wife.

Duperrey, Louis-Isidore
(1786–1865)
French
Hydrographer, Geophysicist

Duperrey was born in Paris on or about October 21, 1786, and is first noted pursuing science in 1809 when he was engaged in the hydrographic charting of the Italian coast. The first ship he commanded in 1814 was the *Coquille*. By 1817, he was chief of hydrography (measuring the depths of coasts as well as charting the shape of the land) on Louis de Freycinet's expedition, from 1817 to 1820, to the South Pacific and New Zealand.

The Earth is not a perfect sphere, and Duperrey's measurements of the small differences in gravity enabled him to chart its shape. In addition, he studied the differences in Earth's magnetism at different locales and mapped the magnetic lines of longitude. (The North Magnetic Pole is not at the geographic North Pole.)

A second oceanic expedition was planned, and Duperrey left France in 1822, commanding the *Coquille* with second-in-command JULES DUMONT D'URVILLE and the naturalists René Lesson and Prosper Garnot. Notably, this 31-month-long expedition came back with all its complement, having lost not a single man—a remarkable feat for sailing ships of the time. They brought back detailed information on ocean currents, terrestrial magnetism, meteorology, and a wealth of specimens destined for the Musée d'histoire naturelle in Paris.

Duperrey's particular interest was the charting of the Earth's magnetic equator. Since the magnetic equator moves over the surface of the Earth, as do the magnetic poles, this is an ongoing occupation. He was one of the earliest researchers to realize this.

Publication of his logs and determinations can be found in *Voyage autour du monde exécuté par ordre du roi . . . 1825–1830* (Travel around the world carried out on the king's orders . . . 1825–1830). This work comprises seven volumes with illustrative plates, and maps. He had been elected a member of the Academie des sciences in 1842 and served as the vice president of that body in 1849 and its president in 1850. Duperrey spent the rest of his life in Paris where he died on August 25, 1865.

E

Earle, Sylvia Alice
(1935–)
American
Biologist, Oceanographer

Also known affectionately as Her Deepness, or Queen of the Deep, Sylvia Earle holds the record for the deepest dives performed by a woman. Born in Gibbstown, New Jersey, on August 30, 1935, Sylvia Earle's family moved to western Florida when she was a teenager. It was there that she began diving, using a hard-hat rig diving suit, and amassed a collection of aquatic plants and animals. Her undergraduate thesis on the algae of the Gulf of Mexico was presented to Florida State University for a degree in botany. Her M.A. in botany is from Duke University where she returned for a doctorate in botany in 1966, after marriage and children. While still a graduate student, Earle took part in a National Science Foundation expedition to the Indian Ocean in 1964. This was one part of the international effort to study the Indian Ocean.

When Earle's first professional dives began, she was one of several scientists using a submersible with a lockout chamber. This device enabled divers to leave the submersible and move around unimpeded. In 1970, she participated in the Tektite Project, sponsored by the U.S. Navy, the Department of the Interior, and the National Aeronautics and Space Administration (NASA), in which scientists lived and worked undersea for extended periods, and during which, the research team experienced an undersea earthquake. In 1977, she undertook a project with a photographer, A. Giddings, tracing the migration paths of whales, which appeared as a documentary film, *Gentle Giants of the Pacific*.

An extension of the earlier deep-diving work culminated in her record dive in 1979, using the JIM suit, an armored piece of equipment designed and made by Graham Hawkes, whom Earle married that year. The firm started by Hawkes and Earle also produced unmanned submersibles. Both types of equipment are used by offshore drilling rigs to inspect the footings of their towers.

In 1990, Earle was appointed chief scientist of the National Oceanic and Atmospheric Administration (NOAA), and in that office, she became involved in the survey and direction of the cleanup after accidental oil spills such as that of the Exxon Valdez, and the deliberate spill by Iraq early in the Persian Gulf War in 1990–91. Resigning her NOAA commission in 1992, Earle has devoted her time to raising public consciousness about the fragility of the ocean biome and its importance to life on Earth. She is presently involved in the direction of several institutions concerned with the exploration of the oceans and the preservation of the oceanic biomes. Earle

The oceanographer Dr. Sylvia Earle speaks during a Capitol Hill Ocean Week (CHOW) event June 6, 2002, on Capitol Hill in Washington, D.C. *(Photo by Alex Wong/Getty Images)*

is now an explorer-in-residence at the National Geographic Society. Some of her publications aimed at the general reader are *Exploring the Deep Frontier* (1980); *Sea Change: A Message of the Oceans* (1995); *Wild Ocean: America's Parks Under the Sea* (1999); and two books for children, *Dive! My Adventures*, and *Hello, Fish.*

⊠ Ehrenberg, Christian Gottfried
(1795–1876)
German
Biologist, Microscopist

This talented biologist was born in Delitzsch, then in Saxony in Germany, on April 19, 1795, the son of Johann Gottfried Ehrenberg, a municipal judge. Sent to boarding school at 14 after his mother, Christiane Becker, died, he first studied to be a clergyman, graduating in 1815. Changing direction, he began to study medicine first at Leipzig and then at Berlin, where he finished his studies in 1817. He received his license in 1818, after producing a doctoral thesis on the fungi of the Berlin region. This work was noticed by the Deutsche Akademie der Naturforscher Leopoldina (Leopoldine German Academy of Natural Scientists), which invited him to be a member. On the recommendation of ALEXANDER VON HUMBOLDT, an internationally known science writer and explorer, the group asked Ehrenberg to be part of a scientific expedition to Egypt, funded by a local nobleman. The group of 10 scientists departed in 1820. They amassed a vast assortment of specimens and drawings of terrestial and aquatic Red Sea organisms. Most of the specimens and nine of the original party were lost in a series of catastrophes, ending in shipwreck. The sole survivor, Ehrenberg, returned to Berlin in December 1825.

By 1827, Ehrenberg was a member of the Berlin Academy of Sciences and a member of the faculty of the University of Berlin. In 1829, Czar Nicholas I of Russia commissioned Humboldt to lead an expedition into Siberia to collect and study botanical and geological specimens. The recently discovered "animalcules" (infusorians, or unicellular organisms) were of special interest. Among the scientists invited were Ehrenberg and the geologist Gustav Rose. This was a relatively short, eight-month-long expedition. After their return, Ehrenberg cemented the connection to Rose by marrying his niece Julie Rose in

1831. Three years after his first wife's death in 1848, Ehrenberg married Karoline Friccius. He spent the rest of his career in Berlin.

The continuing improvements in microscopes made Ehrenberg's work possible. His contribution was in the study of hydrozoa and mollusks, coral polyps, and medusa of the Red Sea, and unicellular organisms. He explored the phenomenon of marine phosphorescence and correctly attributed it to microorganisms. Ehrenberg's body of work included the identification of some rock strata with the fossil remains of microorganisms. This effort established Germany as a country actively involved in the research aspects of micropaleontology and microgeology. One of his guiding principles was that only empirical knowledge, that is, data, can be valid; all else is theory. To tighten the connection to real evidence, he continued to compare specimens from the Red Sea to others from the nearby Baltic, and to fresh samples supplied to him by MATTHEW MAURY, the American navy captain and oceanographer, among others. Maury's samples were taken from depths of 300–365 m (10,000–12,000 ft).

Ehrenberg first began his researches against the background of the work of Gottfried Wilhelm Leibniz, the 17th-century diplomat and polymath who was Isaac Newton's contemporary and mathematical rival. Leibniz attempted to explain everything, including biology, in mechanistic terms. The Leibniz idea of "scale of being" was one means of explaining the differences between organisms. Some scientists still thought that the simplest organisms had arisen from inanimate matter (minerals). Ehrenberg vehemently rejected that theory, known as spontaneous generation. He had discovered that even fungus spores come from a sexual reproductive process: nothing comes from nothing, even the tiniest spore. Ehrenberg also rejected much of GEORGES CUVIER's classifying hierarchies, particularly his collection of medusae, polyps, infusorians, and echinoderms into one heterogeneous category,

which he called Radiata, and dismissed as "lower animals."

While not at all ready to accept CHARLES DARWIN's theory of evolution, Ehrenberg could accept the idea of man as a mammal. There was ample evidence of this, and he did not dispute it. To him there was an unbridgeable gap between man and all the other mammals. Darwin, however, could accept the idea that man was not a special creation and that the similarities to other primates were evidence of kinship.

Ehrenberg also believed that microorganisms possessed the same organ systems that were observed in more complex animals. He argued this point with FÉLIX DUJARDIN, his French counterpart, in a famous series of papers. Now we know he was spectacularly wrong.

Although his work on the Red Sea organisms was never finished, in 1828 Ehrenberg had published part of that effort in *Reisen in Aegypten, Libyen, Nubien und Dongola* (Travels in Egypt, Libya, Nubia, and Dongola). His other large work, published in 1838, was *Die Infusionsthierchen als volkommene Organismen* 1838 (The infusoria as complete organisms). He continued his scientific interests and correspondence until his death on June 27, 1876.

Ekman, Vagn Walfrid
(1874–1954)
Swedish
Oceanographer, Physicist

The youngest son of Frederik Larotz Ekman, Vagn Walfrid Ekman was born in Stockholm, Sweden, on May 3, 1874. He was educated in Stockholm and then at the University in Uppsala in physics with a special emphasis on hydrodynamics and oceanography, his father's field. The subject of Ekman's doctoral dissertation, presented in 1902, was based on VILHELM BJERKNES's "wind spiral." This work was a continuation of the data on "right-hand drift" collected by both Bjerknes and

FRIDTJOF NANSEN. Bjerknes was a famous meteorologist and Nansen an arctic explorer-scientist. Ekman expanded the mathematical calculations and published this work in 1905 as *On the Influence of the Earth's Rotation on Ocean Currents*. In this book, Ekman explained that each layer of seawater is propelled by the layer above it and all move clockwise in the Northern Hemisphere, and counterclockwise in the Southern Hemisphere as a direct result of the Earth's rotation. Each layer down moves more to the right (or left in south latitudes) than the one directly above it. This concept is essential to the understanding of how deep ocean water circulates.

After his assistantship in the Oceanography Laboratory in Oslo (1902–08), Ekman accepted an appointment as lecturer in mechanics and mathematical physics at the University in Lund, Sweden. Two years later, he was made professor of the department and remained at that post for the rest of his working life.

After completing his work on the right-hand drift, Ekman continued to explore the mathematics of wind-driven oceanic circulation. In a test in 1930, he explained the mathematics of a phenomenon known to ship captains as "dead water." This is a phenomenon associated with narrow ocean passages such as the fjords of Norway, where bodies of water are so stratified that moving a ship through them is very difficult. The water itself acts as a brake. Part of his research involved the sampling of water at particular depths, and to improve results, Ekman redesigned the Nansen collecting bottle—the classic tool for sampling. He also designed and produced a current meter with which to accurately measure the velocity and direction of ocean currents. Skillful use of such meteorological information can help mariners avoid zones of dead water. Some of his Atlantic Ocean measurements were lost in World War II but were finally found and published in 1953. Ekman died at his home in Gostad, Sweden, on March 9, 1954, having produced a number of publications in science, philosophy, and religion.

⊠ Ewing, William Maurice
(1906–1974)
American
Geophysicist

The oldest of seven surviving children of Floyd and Hope (Hamilton) Ewing, William Maurice was born on May 12, 1906, in Lockney, Texas. A gifted child, he graduated from high school at 15. His high school mentor persuaded the Rice Institute (now Rice University) in Houston to grant him a scholarship. He graduated with honors in mathematics and physics in 1926, and stayed on to earn an M.A. in 1927, and a Ph.D., both in physics, in 1931. His first publication, *Dewbows by Moonlight*, was published in 1926. Part of his doctoral dissertation was published in 1931 as *Calculation of Ray Paths from Seismic Time Curves*. He accepted a position teaching physics at the University of Pittsburgh in 1929 and a year later went to the physics department at Lehigh University in Bethlehem, Pennsylvania. This position overlapped with a long research leave; he remained officially in the physics department at Lehigh until 1944.

When Ewing presented a paper at a meeting of the American Geophysical Union in 1929, WILLIAM BOWIE (U.S. Coast and Geodetic Survey) and Richard Field (professor of geology at Princeton) suggested that Ewing use his techniques to examine the continental shelf. He did that with HARRY HESS, using a Woods Hole vessel, the *Atlantis*. By 1936, they had traced the gravity anomaly discovered by FELIX VENING MEINESZ. The measurement of gravity on the sea bottom became a lifelong project for Ewing. The gravity of the Earth is predictable since the Earth is a measurable ellipsoid (a body shaped like a football, which, seen side-on, is an ellipse, seen end-on, is a circle). If the gravity measured at a spot on the Earth's surface is not what it is predicted to be, that is a gravity anomaly. In such places, the geology is interesting and complex. Worldwide gravity readings were an issue of particular interest to Vening Meinesz, a Dutch physicist.

Ewing's work schedule left little time for a personal life. His 1928 marriage to Avarilla Hildenbrand ended in 1941. By then, he was a research associate at Woods Hole Oceanographic Institution and working with Allyn Vine and J. L. Worzel, two former students. Together with Columbus O'Donnell Iselin, the director of the institution, they wrote *Sound Transmission in Sea Water*. This described their discovery of a low-velocity sound channel in relatively deep water. This sound fixing and ranging (SOFAR) channel traps sound waves at depths from 700–1,300 m (2,300–4,300 ft) and enables their transmission over great distances. This means that an underwater sound can be "heard" many miles away from its point of origin, an extremely important consideration in wartime.

Ewing had joined the geology department at Columbia University in New York City in 1944, the year of his marriage to Margaret Kidder. A postwar gift to Columbia University of the Lamont estate on the Hudson River meant

William Ewing *(Lamont-Doherty Earth Observatory, Columbia University, courtesy AIP Emilio Segrè Visual Archives)*

that the university could establish a geophysical laboratory and research center there. Ewing subsequently divided his time between the geology department; the Lamont Geological Observatory, where he became director in 1949; and Woods Hole, where he was a research associate from 1950 onward. He carried out research voyages on Woods Hole ships until the Lamont Geological Observatory bought the *Vema*, its own research vessel, in 1953.

Ewing was engrossed in using his seismic refraction techniques to measure the undersea depth of the Moho discontinuity (discovered by ANDRIJA MOHOROVIČIĆ). The discontinuity is a boundary layer separating Earth's crust and the magma layer of the inner Earth. Quickly realizing that the 5 km (~3 mile) average depth of the Moho was less than expected, Ewing and his students looked for a better method. The measurement of the depth of the Moho layer involved bouncing sound waves to it and recording the time required for the returning echo. The estimated reference points came from timing surface waves produced by earthquakes. Together with Frank Press, Ewing studied seismic waves on land and realized that they could be used as a way to measure Moho depths at sea without actually going to sea. Taking the effects of water and sediment into account, Ewing's seismographic studies show that the 5 km depth is correct. This work, published in 1949, showed that the oceanic crustal layer is much thinner than that of the continents—a significant geologic discovery.

As director of the Lamont-Doherty Laboratory, Ewing concentrated on collecting data and sharing it with all the researchers. His efforts built a single large database of oceanographic information that was available to researchers in all fields, not guarded as a private departmental preserve, the more traditional practice. While Ewing did not originally espouse the tectonic plate theory, his excellent collection of data in several fields was largely responsible for upholding this

view of the Earth's conformation. He accepted the idea in 1966.

The use of sound was central to Ewing's work in several interrelated projects. Starting in 1949, and continuing for about 10 years, he worked with his younger brother John Ewing, a geophysicist who also worked at the Lamont Geological Observatory, on sediment maps of the North Atlantic. Both Ewings' work entailed the examination of the contours of the sea bottom. Turbidity currents had been suggested earlier by ARTHUR HOLMES, among others, as an explanation for the formation of deep fissures in the sea bottom, particularly the continental shelves. These undersea currents had been theorized in the 1930s by Reginald Daly, who called them density currents. PHILIP KUENEN later built a laboratory model to show where and how river sediment is deposited, and how these currents cut canyons into it. Kuenen's model was more evidence of how the seafloor was changed and shaped by currents and sediments.

Beginning in the early 1950s, Ewing joined with BRUCE HEEZEN and David Ericson to study the turbidity currents of the North Atlantic. Ewing had discovered the Great Abyssal Plain in 1947. It is a large area covered with sediment that is geologically young; it dates from the Eocene and lies between the Bermuda Rise and the Mid-Ocean Ridge. Two years later, Ewing and his associates traced the length of the Hudson Canyon—the undersea extension of the Hudson River—for 320 km (200 miles) past the edge of the continental shelf and 4,820 m (~16,000 ft) below the surface. The existence of turbidity currents explains the shape and size of the canyon, the plain, and its covering of recent sediments.

Starting in 1952, Ewing concentrated on the magnetic profiles of the seafloor. These yielded further evidence of seafloor spreading from the ridges outward—a vital part of plate tectonic theory. The core sediment samples retrieved by Ewing's expeditions showed the expected magnetic reversals (see FELIX VENING MEINESZ,

DRUMMOND MATTHEWS, and FREDERICK VINE) that are further confirmation of the theory of seafloor spreading and the production of new edges to the crustal plates.

Ewing's second marriage ended in 1965, and he married Harriet Bassett that year. When the University of Texas at Galveston made him an offer in 1972, he moved there and was professor of geology until his death of a cerebral hemorrhage on May 4, 1974. His wife donated his papers to the university archive.

Ewing produced an enormous body of work. He was a member and officer of a number of scientific societies, including the American Geophysical Union, Sigma Xi, and the Seismological Society of America. He served on the advisory committees of the National Science Foundation and the International Geophysical Year and was the recipient of numerous medals from (among others) the National Academy of Sciences, the American Geophysical Union, and the U.S. Navy.

F

Feely, Richard Alan
(1947–   )
American
Chemist

Richard Feely was born far from the sea in Farmington, Minnesota, on February 26, 1947. He finished his undergraduate work in 1969 at the College of St. Thomas and went on for an M.S. degree in chemistry at Texas A & M (1971) and a Ph.D. in chemistry in 1974. His present positions at the Pacific Marine Environment Laboratory (PMEL), a branch of the National Oceanic and Atmospheric Administration (NOAA), as well as in the department of chemical oceanography at the University of Washington, are the sites for his research into the chemistry of seawater around black smokers.

This field of research involves studying the factors influencing the trace elements in marine particulates and sediments around the hypothermal vents. He has worked with KAREN VON DAMM on the plumes of hot water rising from the Pacific floor near the Juan de Fuca plate—an area of frequent volcanic activity off the coast of Washington and British Columbia. As the very hot water rises, it affects the chemistry of the surrounding ocean water because it has carried iron, manganese, and sulfur upward, which oxidize in the cold and relatively oxygen-rich water. Vanadium,

arsenic, chromium, and phosphorus are carried up along with the chemosynthetic bacteria that live in the hot spots beneath the ocean floor. These bacteria maintain their energy balance by chemical reactions that are not related to photosynthesis. They may be the relatives of the earliest life-forms on Earth and may also be present on other solar system bodies; this idea, however, has not been proved conclusively. Some researchers believe that the primitive iron-sulfur bacteria are of more recent origin. In any case, they do exist in chemical environments that are extreme both in temperature and mineral content.

Feely is an active researcher. His work deals mainly with marine environments; however the estuarine and riverine systems are not ignored. Most of his publications concern the metallic content of water and appeared in journals such as *Journal of Geophysical Research*, *Proceedings of the National Academy of Sciences of the United States*, *Deep Sea Research*, and *Marine Chemistry*.

Ferrel, William
(1817–1891)
American
Geophysicist

Born January 29, 1817, in a farming community in Bedford County, in southern Pennsylvania, Ferrel was the eldest of Benjamin Ferrel's eight children. The family moved to Virginia (now West Virginia) in 1829, and Ferrel had several years of classes in winter in a one-room school. In 1832 he observed a partial solar eclipse and was so fascinated by it that he used an almanac and a geography book to teach himself how to predict eclipses. While still an adolescent, he taught school for several years to earn money for further education and graduated from the newly created Bethany College in 1844. He continued to teach and began to seriously study Isaac Newton's *Principia,* the basic treatise on calculus, and NATHANIEL BOWDITCH's translation of *Mécanique Céleste* (Celestial mechanics) written by the French astronomer, Pierre-Simon, marquis de Laplace. The first paper Ferrel wrote, published in 1853, established his scientific reputation. He gave up his teaching career in 1857 when he was appointed to the staff of the American Ephemeris and Nautical Almanac. While there, Harvard mathematician and superintendent of the U.S. Coast Survey Benjamin Pierce persuaded Ferrel to join the Survey. This led to another government position in 1882, when Ferrel joined the U.S. Army Signal Service, the forerunner of the Weather Bureau.

Ferrel's life's work grew from the theoretical works of Newton, DANIEL BERNOULLI and Leonhard Euler on the action of the Moon and Sun on the ocean's tides. In his first major work, he explored the possibility that the tides retard the Earth's rotational velocity. Earlier, Laplace had asserted that there was no appreciable effect, but Ferrel demonstrated that the effect was not negligible and that it accounted for the lunar acceleration, in which the distance between the Earth and the Moon is slowly increasing because of tidal friction. Laplace had also stated that diurnal tides (two high tides and two low tides daily) would disappear if the ocean were all at a uniform depth; Ferrel presented a work that disputed that conclusion as well.

The study of tides was Ferrel's major interest. Ancillary works that grew out of this central theme were a study of shallow-water tidal factors, the calculation of the mass of the Moon, and the construction of the tide prediction machine. This device, a primitive analog computer, was operational by 1883. Ferrel continued to study the mathematics of Earth's rotation and the effect of rotation on the motion of bodies on the surface—either land or water—and the consequent effect on meteorology. His meteorological studies were aimed at trying to find a cause for the variation in air pressure. MATTHEW MAURY, a naval captain studying tides, weather, and depth soundings, had collected data in regions of high pressure (medium and high latitudes) and also in low-pressure areas (the equatorial region). Ferrel wrote "An Essay on the Winds and Currents of the Oceans" using Maury's data, which was published in the *Nashville Journal* in 1856.

In 1861, the *American Journal of Science* published Ferrel's essay "The Motion in Air." This work, later expanded and reissued as a pamphlet, mathematically explains the formation of gyres (giant circular surface currents) in the ocean. This explanation is generally known as Ferrel's Law; a body in motion is deflected to the right in the Northern Hemisphere and to the left in the Southern Hemisphere as a result of the rotation of Earth. Ferrel put American meteorology on a firm scientific and mathematical basis. His work led directly to science-based and government-supported weather forecasting stations.

Retired at 70, Ferrel moved to his brother's farm in Maywood, Kansas; he found it dull and returned to West Virginia. Old and frail, he found that he could not manage alone and returned to his brother. He died in Maywood, Kansas, on September 18, 1891.

⊠ **Forbes, Edward**
(1815–1854)
English
Invertebrate Zoologist, Paleontologist

Edward Forbes, born on February 12, 1815, on the Isle of Man, was the eldest of the eight surviving children of a paper manufacturer turned banker, also named Edward, and his wife, Jane Teare. A younger brother, David, was a well-known geologist. Edward Forbes was educated in local schools until 1831, when he went to London to study art. He did not think he was as good as his competition and, following his mother's wish, he then went to Edinburgh for medical training. While there, he met and worked with Robert Graham, the botanist, and Robert Jameson, a noted zoologist. A strong interest in marine organisms and their distribution led Forbes to travel to Norway in 1833. The account of this voyage was published in 1835. In 1837 he wrote *Comparative Elevation of Testacea in the Alps* that was a first step toward the definition of biomes (biological communities). After his mother's death in 1836, Forbes left the British Isles for Paris and the biological lectures of Henri Blainville and ÉTIENNE GEOFFROY SAINT-HILAIRE.

By 1839, he had published five papers on mollusks. He followed this with a definitive work on British echinoderms, spiny invertebrates typified by the starfish. In this work, which appeared serially between 1839 and 1841, Forbes discussed echinoderm distribution, the relation of families of these animals to their location, and the relationships of living species to fossil species.

Tired of living on the occasional lecture on natural science, he joined a voyage to the Aegean Sea in 1841. As the naturalist on board, Forbes was actively involved in systematic dredging and cataloging of the organisms found in the area, both living and fossil. This was the first scientific examination of the eastern Mediterranean since Aristotle's work there. Unfortunately, Forbes contracted malaria while in the Aegean and received the news that his yearly stipend from his father had ended; also he needed to find employment.

On his return to England, Forbes was appointed to the chair of botany at Kings College in London in October 1842, as well as curator at the Geological Society. This position involved managing the society's collection and library,

and editing its journal. In spite of illness, Forbes continued to work and also published *On the Mollusks and Radiata of the Aegean Sea . . .* for the British Association for the Advancement of Science. The work on biotic communities continued, and Forbes, working with the evidence that the numbers of species declines with increasing depth, extrapolated that there existed an "azoic zone" where no plant or animal species lived, at 300 fathoms or about 550 m (1,800 ft) in depth. Forbes was totally wrong in this assumption, as many scientists have proved again and again.

In 1844, Forbes began teaching paleontology for the British Geological Survey in addition to his other duties. He was elected a fellow of the Royal Society in 1845. His pivotal paper, published as "On the Connexion Between the Distribution of the Existing Fauna and Flora of the British Isles and the Geological Changes Which Have Affected Their Area" (1846), incorporated the then-recent discoveries in geology and the acceptance of the effects of glaciation on the European continent. Forbes began the four-volume "History of British Mollusca" in 1848, the year of his marriage to Emily Ashworth, and it appeared in parts until 1852. Forbes was awarded an endowed chair; he was Regius Professor of Natural History in Edinburgh in 1854. Unhappily, the honor was short-lived: he died of what was then diagnosed as kidney failure on November 18, 1854.

Throughout his works, Forbes sought to point out "God's plan" in nature. He was an antievolutionist who believed that the Paleozoic world was replaced by the "Neozoic." Another large work, "On the Tertiary Fluvio-Marine Formations of the Isle of Wight," was completed by his colleagues, Henry Bristow and ROBERT GODWIN-AUSTEN, in 1856. Godwin-Austen also finished the "Natural History of European Seas" in 1859; Forbes had completed the first five chapters of this first general study of oceanography. The naturalist CHARLES DARWIN cited this work extensively in his own writings.

Forchhammer, Johann Georg
(1794–1865)
Danish
Chemist, Geologist

Johann Forchhammer was born on July 26, 1794, in Husam, Denmark. His parents were Johann Ludolph and Margarete Wiggers Forchhammer. His father was a teacher and manager of a teachers college in Husam and then in Tønder. Johann Georg Forchhammer studied in both towns but had his education cut short when his father died in 1810. The boy worked in a pharmacy for five years before entering the University of Kiel for physical sciences, mathematics, and mineralogy. Shortly thereafter, he joined a research group; Hans Ørsted, the famous physicist who established the relationship between electricity and magnetism, was also in this group. Forchhammer worked on the exploitation of coal and ore deposits around the Baltic Sea as part of his education. He presented his doctoral thesis at the University of Copenhagen in 1820 and spent the rest of his life in that city.

A trip early in his career took Forchhammer to England, where he again explored coal deposits, and the publication of his report of this work insured his election to the Royal Danish Academy of Sciences. His career at the University of Copenhagen began with the post of lecturer in geology in 1821, then, in the same year, professor of chemistry at the Polytechnic Institute and manager of the chemistry laboratory there. Once he became the director of the institute, that was the principal post he held until he died. The other lifelong positions Forchhammer held were professor of mineralogy and geology at the university, beginning in 1831, and secretary of the Royal Danish Academy of Sciences from 1851 onward.

Forchhammer's work in marine science began in 1843 when he studied seawater with a geologist's eye, examining the mineral deposits and their relationships to volcanic activity. The samples he tested were collected by British and Danish naval vessels. The results of these investigations showed that while the salt content varies with the sample, the dissolved materials are essentially constant in spite of the addition of some substances—calcium and silicon in particular—that are most plentiful in river water. Forchhammer concluded that the composition of the seawater was not dependent on what the rivers brought into it, but that the dissolved materials in seawater were the remnants left after various substances had been removed. The geochemical balance in seawater was low in calcium and silicon because those elements had become incorporated into nonsoluble materials.

Forchhammer was called the "father of Danish geology" as early as 1835. He was certainly a huge influence on those who followed him in this field. Forchhammer died in Copenhagen on December 14, 1865.

Forster, (Johann) Georg Adam
(1754–1794)
German/English
Naturalist, Geographer

The eldest son of JOHANN REINHOLD FORSTER and Justina Forster, Johann Georg Adam Forster was born on January 17, 1754, near Danzig when his father was a clergyman there. He was considered a child prodigy and was quite gifted in languages, as was his father. Aware of their child's abilities, his parents educated him at home until he left for England with his father in 1766. The younger Forster, then 11, studied at the Dissenters Academy and also taught some language classes and helped his father to translate LOUIS-ANTOINE DE BOUGAINVILLE's account of his voyage around the world in 1766–69.

Forster was the assistant artist on JAMES COOK's second voyage to the Pacific Ocean in 1722–76, and much of the account of this voyage by both Forsters (his father served as the ship's naturalist) was written by Georg. Produced in defiance of the admiralty's requirements, this is more a literary work—a forerunner of VON HUMBOLDT's work—than a scientific document. It did serve to move his supporters to name Georg a fellow of the Royal Society in 1777. Because of the dispute with the admiralty, however, Forster could find no position in England. When he and his father returned to Germany separately, Georg was appointed professor of natural history at the University in Kassel in 1779. He moved to the University in Vilna, then in Russia, now Lithuania, six years later, and married Thérèse Heyne that year. In 1786, he published a work based on his medical dissertation in which he described the flora of Australia. By 1787, Forster was back in Germany as librarian in the University of Mainz. Humboldt was then in Paris and planning his own voyages of exploration. He invited Forster to accompany him to England. This successful trip left Forster ready for another adventure. Since he thought of himself as a sociologist, he was attending the legislative National Convention in Paris, observing this new phenomenon of government as democracy, when he died suddenly on January 10, 1794. Forster was a charming man who literally could do anything. His travel accounts were perhaps the first scientific travelogues—a genre that was continued by Humboldt, Beebe, Cousteau, and others.

Forster, Johann Reinhold
(1729–1798)
German/English
Naturalist

Perhaps Forster should be described as an English-German naturalist. His family had originally come from Britain, where they were Royalists in the Civil War, and fled to the area near Danzig (now Gdansk, Poland). John (or Johann), the son of Georg, was born on October 22, 1729, and educated as a clergyman at Marienwerder and Berlin in 1748. He then was a pastor in Halle and three years later led a parish near Danzig. His interests were not in theology, and given the chance, he became a bureaucrat working for the government of Empress Catherine the Great of Russia. During his 12 years of working in Russia, he maintained a correspondence with scientists such as Linnaeus and Joseph Banks. In 1765, Forster led an expedition to explore the lower reaches of the Volga River. He did not receive an expected promotion, and in 1765 went to England with his son JOHANN GEORG ADAM FORSTER (known as Georg), who was then 11 years old, where the elder Forster succeeded Joseph Priestley as teacher of natural history in Warrington, the Dissenters Academy, in Lancashire.

On the request of the Royal Society, Forster cataloged and annotated the Hudson's Bay Company's collection of specimens. Published in the Philosophical Transactions of the Royal Society, this work is the first scientific treatise on North American zoology. A work for the same publication on minerals and then another work on British insects followed.

In 1771 Forster translated the journal of the navigator LOUIS-ANTOINE DE BOUGAINVILLE, who had commanded the first French expedition around the world in 1766–69. The following year, the HMS *Endeavor* returned from its first Pacific voyage under JAMES COOK's command. Forster was determined to accompany Joseph Banks, the naturalist, on the next Pacific voyage. However, Banks did not go; his requests for accommodation were unmet by the admiralty. As a replacement, Banks suggested that Forster accompany Cook. Since Forster had been elected a fellow of the Royal Society in 1772, he was a suitable candidate. The first sign that he might be a troublesome shipmate was his insistence on taking his son Georg along.

After the voyage ended in 1775, Forster published a preliminary work largely on the botanical specimens he brought back. Later, both Forster and Daniel Solander, who had accompanied Banks on Cook's first voyage, each prepared a more complete catalog. Forster's was *Observations Made During a Voyage Round the World*, published in 1778. In it, he described the geography, natural history, and what might now be called the sociology of the regions visited. The first detailed, scientific observations of corals and earthquakes were in Forster's account, and these were of great interest and inspiration for CHARLES DARWIN, who followed this path 60 years later.

By 1780 the Forsters were penniless—the book did not sell, and they had fought with the admiralty and the scientific community. Father and son returned separately to Germany. In 1787, the elder Forster accepted a position at the University of Halle as professor of natural history, mineralogy, and medicine that he held until his death on December 9, 1798.

⊠ Franklin, Benjamin
(1706–1790)
American
Physicist, Meteorologist, Oceanographer

Benjamin Franklin, an unquestioned American genius, was born in Boston on January 17, 1706. His parents were Josiah Franklin and his second wife, Abiah Franklin (also known as Jane), who had a very large family. After an early start in the printing business with his brother, Benjamin Franklin broke his indenture (a contract for employment and housing) to his half brother James, and moved to Philadelphia in 1723. He was associated with that city for the rest of his eventful life. He was an intensely curious man, and marine science was, for Franklin, a tangential interest. His major scientific research was concentrated on the study of electricity, a fashionable curiosity of the day.

As the state's representative to the British Crown, he began a series of ocean voyages—eight

Benjamin Franklin *(E. F. Smith Collection, Rare Book & Manuscript Library, University of Pennsylvania)*

in all—that built on his early fascination with the sea. As Postmaster-General of the colonies, Franklin had the first chart of the Gulf Stream printed in 1770 after about 20 years of interrupted study. The impetus for this map grew out of the complaints of the Boston Board of Customs: mail from England took two weeks longer to travel to Boston than did merchantmen sailing from Boston to England. Talking to Nantucket whalers, Franklin found that they knew "the Stream" because whales were found, not in it, but on its borders. Captain Timothy Folger drew the map that was then engraved and printed by the post office.

One series of Franklin's experiments involved careful use of thermometers. These fragile instruments were expensive equipment at the time. On his voyages between America and England in the 1750s and 1760s, Franklin recorded the slight differences in water temperature and used them as a navigational tool, attempting to correlate the variation with increasing depth. He was moderately successful.

After studying cloud formation at sea, Franklin theorized that convection currents existed in air. The idea that air was not uniform everywhere was a new concept. Franklin's most spectacular and large-scale experiment that might be called marine science was his demonstration for the Royal Society at Portsmouth Harbor, England's great naval base on the southeast coast. The details of this demonstration were undated in his diaries, but it was during one of his long stays as the agent of the Pennsylvania colony, probably between 1757 and 1762. He attempted to show that "pouring oil on troubled waters" really does smooth the waves, but it is most effective in smaller areas.

All science, for Franklin, was an exercise for his mind: it was logical, and the results, while they could be surprising, did have a discernable cause. He managed to make significant contributions to man's understanding of physical phenomena, while fully occupied as America's statesman-savant. His last public act was taking part in the Constitutional Convention in 1787–1788, and he died in Philadelphia on April 17, 1790.

G

⊠ **Geoffroy Saint-Hilaire, Étienne**
(1772–1844)
French
*Zoologist, Comparative Anatomist,
Natural Scientist*

A number of notable Geoffroys have contributed to science. Étienne was born on April 15, 1772, in Étampes, near Paris, the youngest of 14 children. After receiving a degree in law in Paris in 1790, he continued his studies in science and medicine at the Collège du Cardinal Lemoine.

That was a dangerous time in Paris. The Revolution, which had started as an attempt to reform government on the American model, had become increasingly violent. Geoffroy's activities included helping his friend, René-Just Haüy, avoid the guillotine because he was a priest. Haüy survived the Terror and went on to a distinguished career in mineralogy. By 1793, Geoffroy was appointed professor of vertebrate zoology at the former Jardin du roi, the royal research center, later renamed Jardin des plantes, which quickly became the Musée national d'histoire naturelle after BERNARD LACÉPÈDE, the curator, fled the Revolution. A recommendation by Henri Tessier, a fellow zoologist, led Geoffroy to invite GEORGES CUVIER to Paris as his coworker, later his rival.

When Napoléon invaded Egypt in 1798, Geoffroy was one of the accompanying naturalists. Napoléon planned a grand expedition to Egypt in the style of Alexander the Great. Conquest was only part of his plan. He invited many scientists, engineers, linguists, and philosophers to accompany the army and conduct serious studies of every aspect of Egyptian life and history. Cuvier refused the invitation, but Geoffroy Saint-Hilaire accepted. He saw it as an opportunity to further his career through a personal association with Bonaparte. He sent trunks full of specimens back to Paris for study. On examining the mummified cats and birds, Cuvier declared that evolution was impossible since these specimens were no different from cats and birds alive in Paris. After his return, Geoffroy resumed his position in the Jardin des plantes. He was then elected to the Académie des sciences in 1807, and in 1809 he was named professor of zoology at the University of Paris, a post he held officially until he died.

The great debate in biology in the early 1830s was centered on vertebrates. The vast array of invertebrates was not well known. Biology was then a concentrated study of those animals that were familiar to the researchers. There was intense argument about structure, function, and form. This study was later named morphology. Cuvier argued for functionalism—if animal structures did the same task, the animals were related; Geoffroy Saint-Hilaire followed the theories of GEORGE BUFFON and JEAN-BAPTISTE LAMARCK more formally, that all vertebrates were modifications of a single basic form or archetype. Geoffroy looked for homologous structures, but he did not make the leap from the vertebrate to the invertebrate body plan. Homologous structure is a concept in comparative anatomy. On dissection, it is easy to see that the foreleg of a horse, a human arm, and a bird's wing have similar, but not identical, bones. Each of these sets of bones also has similar musculature attached. Geoffroy saw them as similar and went on to equate insect shells to ribs and vertebrae.

When in 1830, two naturalists, Meyranx and Laurencet, in discussing cephalopods (squid, octopi, and cuttlefish), drew comparisons to vertebrates. As increasing numbers of invertebrates became known, zoologists who were well-versed in the anatomy of vertebrates were determined to find homology. The newly discovered animals were seen in terms of the body plan of those creatures that were already well known, tetrapod (four-legged) vertebrates. Cuvier could not accept this and challenged Geoffroy Saint-Hilaire, the originator of this theory, to a series of eight public debates, held between February and April of 1830. In spite of this public airing of differing views, the two did not become enemies. The end result of the debates gave Cuvier a slight edge. He did not overstate his case.

Geoffroy's later work was on unity of type. This was his attempt to explain that in performing basic metabolic functions, an animal must have certain structures. Thus, if vertebrates need lungs to breathe, invertebrates also need to breathe, and thus, they must have "breathing structures," organs that do what lungs do in a vertebrate body. This idea fell easily into evolutionary theory, which Geoffroy never did espouse, but he was tending in that direction when, in 1840, an incapacitating stroke left him blind and disabled.

The split between Geoffroy Saint-Hilaire and Cuvier was typical of the biology of the day. Both were comparative anatomists and truly morphologists. Comparative anatomy focuses on similar or analogous structures and organs. Morphology describes how they look and function. Geoffroy's work was used extensively by later biologists, notably CHARLES DARWIN, who was much taken with the idea of homologous structures. Late in his career, Geoffroy wrote of changes in structure that would lead to improved chances of survival. Unlike Lamarck, he did not believe that these were slow, gradual transformations but rapid changes in embryology. And unlike Darwin, he did not organize these ideas into a theory. His interest in embryology led Geoffroy to be the first to scientifically examine teratology, and his article "Monstre" in the *Dictionnaire classique d'histoire naturelle* (Classic dictionary of natural history) is a significant first step. This early discussion of congenital malformations (two-headed calves or children with cleft palates) treated these as the result of some cause other than the act of a vengeful god. He followed that article with a larger work on the subject, *Histoire . . . des anomalies d'organization chez l'homme et les animaux* (Organizational anomalies in man and animals) in 1832–37. Like Cuvier, he was sure of the importance of paleontological evidence; and in spite of their errors, both Cuvier and Geoffroy each contributed to the significant work of later scientists. Geoffroy died in Paris on June 19, 1844. His son Isidore, who was a zoologist and embryologist, succeeded him as professor of zoology.

⊠　**Godwin-Austen, Robert Alfred Cloyne**
(1808–1884)
English
Geologist, Natural Historian

Robert Austen was born in Guildford, England, on March 17, 1808, son of Henry and Anne

(Bate) Austen. Educated in France, he returned to England and received his B.A. from Oxford in 1830. While at Oxford he became interested in geology because of William Buckland, his teacher, and an exponent of the theory of slow, uniform changes in the Earth's appearance—a uniformitarian. His friends and students included CHARLES LYELL, Leonard Horner, and Roderick Murchison. Austen continued his education in the law at Lincoln's Inn and was sponsored by his geology friends for membership in the Geological Society of London. For several years after his time at Oxford, Godwin served as a circuit judge.

After his marriage to Maria Godwin in 1833, Austen added Godwin to his name. Maria was the only child and heiress of Sir Henry Godwin, a general who had served in Burma. After 1840, Godwin-Austen worked solely as a geologist. Returning to Guildford to the gracious life of a country gentleman, he studied the geology of Surrey and carefully mapped the successive foldings and erosive processes that built the European landmass. For example, he explained the nature and extent of the continuous coal belt that was found underlying vast stretches in England, Belgium, and northern France. This coal bed was an exciting find; the rapidly growing industries of all of western Europe were totally dependent on the availability of a reliable fuel source. Until the discovery of North Sea oil, this coal deposit supplied the fuel requirements of all three nations. Godwin's work was proved to be true by examination of test borings made in the last quarter of the 19th century.

Godwin-Austen was one of the first marine geologists whose primary interest was the geologic history of the English Channel. The early death of his friend EDWARD FORBES left an ambitious work unfinished. Godwin-Austen devoted years to finishing the encyclopedic work Forbes had begun, *The Natural History of European Seas*, published in 1859. In the same year Godwin-Austen completed another work that Forbes had begun, *On the Fluvio-Marine Tertiary Strata of the Isle of Wight*. These works on the fossil-bearing layers of English bedrock were the beginning of a study of the animals of the Tertiary seas that led him to his concentrated study of the structure of the English Channel. Godwin died in Guildford on November 25, 1884.

⊠ **Goethe, Johann Wolfgang von**
(1749–1832)
German
Natural Philosopher

Goethe, son of Johann Kaspar Goethe, a lawyer, and Katherine Elisabeth Textor, the daughter of the mayor of Frankfurt, was born in Frankfurt, Germany, on August 27, 1749, and educated at home until his father insisted that he attend the University of Leipzig to study law. He joined the court at Weimar and stayed there for his lifetime. A prodigious writer and traveler, though not a scientist, he was a pivotal influence on many fields in his own time.

He created the term morphology, which in biology is a descriptive categorization of organisms, but to Goethe in 1817, it meant the study of all sorts of formation and transformation. He included in this chain of being rocks, clouds, colors, living things, and societies. Two influential works he wrote are *Die Metamorphose der Pflanzen* (The metamorphosis of plants) in 1798 *and Die Metamorphose der Tiere* (The metamorphosis of animals), in 1806. The two long poems on metamorphosis were important because Goethe was an internationally known polymath and intellectual star. To his contemporaries, everything he wrote was noteworthy: so if he declared that there had to be a prototype plant and all plants were based on that model, the idea was given serious consideration by botanists. Similarly, comparative anatomists expended much effort in finding the archetypal animal structures, totally convinced that such bodies had to exist. To Goethe, his archetypes did not imply evolution:

Johann Goethe *(Hulton/Archive/Getty Images)*

he theorized that a basic plan had to exist; one had only to look for it and find it. Goethe was not just a writer of plays, novels, travel literature, and poetry. Both poems had a great effect on a number of early biologists: GEORGES CUVIER, FÉLIX DUJARDIN, CHRISTIAN EHRENBERG, ÉTIENNE GEOFFROY SAINT-HILAIRE, ERNST HAECKEL, and others who tried—fruitlessly—to find his archetypes in plants and animals. According to this theory, all organisms were variations of a single type and each type was made of parts that were morphologically identical.

As a scientist (he made that claim for himself), Goethe was an impediment. He is important to the history of science because so much mid to late 19th-century effort was expended in trying to find his archetypal specimens. In another provocative work, *Zur Farbenlehre* (The study of

color), published in 1810, he argued for a study of color that was more than the mere Newtonian theory of angles and light rays; he ascribed emotional properties to colors, leaving empirical science behind. Goethe retired in Weimar, Germany, and died there on March 22, 1832.

⊠ Goodrich, Edward Stephen
(1868–1946)
English
Comparative Anatomist, Embryologist, Paleontologist

Goodrich was born on June 21, 1868, in Weston-super-Mare in southwestern England and grew up in France. His father, the Reverend Octavius Goodrich, died when Edward Goodrich was an infant, and his widowed mother, Frances, took her three children to Pau, France, where they could live affordably. Goodrich returned to England in 1888 to go to University College in London, determined to become an artist. At University College, he met E. RAY LANKESTER, the formidable zoology professor who persuaded him to study zoology instead. When Lankester went to Oxford in 1892 to teach comparative anatomy, he took Goodrich with him as an artist-assistant.

This was a very productive period in Goodrich's career. Some of his publications are *Cyclostomes and Fishes* (1909), and *Evolution of Living Organisms* (1912). The latter was reissued after considerable expansion as *Evolution of Living Organisms: An Account of Their Origins and Evolution* (1921). Rising rapidly in the department, Goodrich was named Linacre professor of comparative anatomy by 1921 and remained in that post until he retired in 1945. In 1930, he published *Studies on the Structure and Development of Vertebrates*. Goodrich's principal research interest was the nephridia of invertebrates. These structures are excretory organs; their analogs in mammals are the kidneys. Concentrating on marine organisms from places as varied as southern England and the Channel, southern Italy, Sweden, Bermuda, Madeira, the Canary Islands, and Southeast Asia, he examined invertebrates, looking for nephridia. These are a product of the outermost layer of a developing embryo. Structurally, they are very different from coelom ducts that arise from the embryonic middle layer and grow outward to form germ cells. In vertebrates, the proto-germ cells also take on excretory function, and the nephridia are lost. The construction of the invertebrate excretory organs is not like that in vertebrates. In invertebrates, the nephridia develop from the same inner layer of the gastrula—the enfolded hollow ball of cells—as does the digestive tract. In vertebrates, the proto-kidney and the reproductive system arise from the same outer layer of the gastrula. This difference appears in the very early embryo.

Another interest that occupied Goodrich for years was the careful differentiation of fish scale structures. The scales in fossil fish are not constructed as are those in modern fish. This data is used in dating geologic strata and determination of the range of fish in earlier geologic times.

Always a popular instructor, Goodrich illustrated his lectures with his own drawings. At Columbia, for example, students came prepared to photograph his blackboards. He was elected a fellow of the New York Academy of Sciences and the Royal Society of London in 1905, and the society's medallist in 1936. Other institutions awarded him honorary doctorates, and he was referred to as the great comparative anatomist of the 20th century. Goodrich died in Oxford on January 6, 1946.

H

Haeckel, Ernst Heinrich Philipp August
(1834–1919)
German
Biologist

Ernst Haeckel was fortunate in that his father, Carl, was an educational bureaucrat married to Charlotte Sethe, the daughter of a privy councillor of Prussia. Born in Potsdam on February 16, 1834, the future scientist was educated at the Domsgymnasium (Cathedral school), and graduated in 1852. His first scientific love was botany, but his parents wished him to study medicine. A medical degree was his path to a scientific career, so Haeckel attended lectures in Berlin, Würzburg, and Vienna, receiving his degree from the University of Berlin in 1857 and a license to practice medicine in 1858. After reading the translation of CHARLES DARWIN's *Origin of Species*, Haeckel was so enthralled that he resolved to abandon medicine as a career and return to his earliest interest, biology. Three years later, he was lecturing on comparative anatomy for the Faculty of Medicine in Jena. The next year, Haeckel was appointed associate professor of zoology at the Faculty of Philosophy there, and rose to the professorship and directorate of the Zoological Institute in 1865. He kept both positions until his retirement in 1909.

Scientifically, Haeckel was a world-renowned figure for many years. During his student days in Berlin, he was introduced to zoology by JOHANNES MÜLLER, who collected Baltic Sea specimens. This led to Haeckel's first scientific publication, and his doctoral dissertation, in 1857, *Über die Gewebe de Flusskrebses* (On the tissues of crayfish). In the years between Haeckel's medical retirement and his arrival at Jena (1858–61), he had participated in a Mediterranean cruise that followed up on Müller's work. The result yielded 144 newly described radiolarian species, published in *Die Radiolarien* (The radiolaria) in 1862. This was the beginning of his work on radiolaria in which he went on to describe more than 4,000 new species. These unicellular radially symmetric plankton continued to interest Haeckel for years. His major book was the two-volume *Natürliche Schöpfungs Geschichte* (Natural history of creation), published in 1868. His contribution to embryology was marked by numerous articles on gastrulation—the enfolding of the developing egg. Several words basic to zoology, such as phylum, phylogeny, and ecology were created by Haeckel: a man who produced much sound science and some very misguided opinion.

He was an enthusiastic scientist and teacher, a member of the Leopoldine Academy of Berlin (1870), the Imperial Academy of Science of Vienna (1872), the Royal Swedish Academy of

Science (1882), the American Philosophical Society (1885), and the Royal Society of Edinburgh (1888), as well as the recipient of many scientific honors.

Influenced by a tradition of philosophy based on JOHANN GOETHE's romantic writings, Haeckel tried to stress spirit and unity of mind and matter, an idea he called "monism." He was convinced that the embryological stages of organisms were derived from predecessors in their ancestral tree—the catch-phrase attributed to him is "ontogeny recapitulates phylogeny." This insistence on a philosophical driving force in science and his intense nationalism, his involvement in his last years with social Darwinism, and racism made him the favorite of eugenicists and later, Nazi propagandists. Haeckel died in Jena on August 9, 1919.

⊠ Haldane, John Scott
(1860–1936)
English
Physiologist

Part of a famous family, John Scott Haldane, son of Robert Haldane, a solicitor, and Mary Burden, whose brother was the Waynflete professor of physiology at Oxford University, was born on May 3, 1860, in Edinburgh, Scotland. He attended the university there, graduating in 1884 with a degree in medicine. The first work he published, in 1887, dealt with air quality and its circulation in houses and schools. His uncle found a post for him in Oxford, where he spent the rest of his working career.

A particularly serious mine accident had made the study of carbon monoxide levels essential. Continuing his work on air circulation, Haldane next looked at mines, concentrating on working conditions, carbon monoxide levels, and causes of death. He developed better methods and instruments for the analysis of blood gases as a result. The papers Haldane published using this

British physiologist Dr. John Scott Haldane, using apparatus for penetrating a noxious atmosphere. *(Hulton/Archive/Getty Images)*

work are "The Action of Carbonic Acid to Oxygen Tension," published in *Journal of Physiology* in 1895, and *Cause of Death in Colliery Explosions*, London, 1896. Haldane's definitive work, a group of papers published in the *Journal of Hygiene* and coauthored by J. G. Priestley, his long-time associate, was on the control of respiration by the partial pressure of carbon dioxide in the blood and the amount of carbon dioxide that is in inhaled air (1905). This work on respiration was continually modified and expanded. The concepts in this study were extended to work on respiration in other stressful conditions—in heatstroke, mines, and diving equipment. The latter led to a design of methods for gradual decompression and study of the reactions of hemoglobin and oxygen, since they affect the removal of carbon dioxide from body tissues for release by the lungs. Haldane's published research papers that deal with undersea work are "The Prevention of Compressed Air Disease," published in *Journal of Hygiene* in 1908; and "The Absorption and Dissociation of Carbon Dioxide by Human Blood," in *Journal of Physiology* in 1914. In 1916, Haldale presented the Stillman lectures at Yale. This series of talks was incorporated into RESPIRATION, published in 1922, and coauthored with J. G. Priestley.

Haldane's principal achievement in public health was the design of ventilating systems for mines, tunnels, ships, diving equipment, and submarines. Its effect on public health was considerable. This study was not merely a philanthropic wish for fresh air in factories, mines, tunnels, and ship interiors. The lives and health of people working in such enclosures for long periods of time was immediately affected if such places were poorly ventilated. Standards of air quality were developed. Air pumps were mandated for installations where normal air circulation is difficult or impossible. The result of forcing air into places with poor air circulation was crucial in the development of diving equipment.

Honors came from the Royal Society of London, which elected him a Fellow in 1897. He received the society's medal, awarded annually for research, in 1916, and the Copley Medal in 1934. He was a Companion of Honor in Industrial Hygiene, with a royal commendation from King George V in 1928, and a Fellow of New College at Oxford from 1901 until his death. Haldane's classic works are "Regulation of the Lung Ventilation," which appeared in the *Journal of Physiology* in 1905, and the aforementioned "Prevention of Compressed Air Disease." These papers were instrumental in assessing the length of time needed for a diver to rise to the surface while avoiding decompression sickness, known as the bends. This is the result of increased nitrogen dissolving in blood at increased pressure. When the diver rises, and the pressure lessens, the nitrogen literally boils out of the blood into joints and body cavities. The result is intense pain and, in severe cases, permanent crippling or death. Every diver, in a pressurized compartment or not, is subject to this danger. Haldane died in his Oxford home on March 15, 1936.

Another Haldane also had a scientific career. J. B. S. Haldane, the geneticist, was John Scott Haldane's son and the subject of several respiration experiments.

⊠   **Halley, Edmond**
(1656–1743)
British
Astronomer, Geophysicist

Born near London on January 14, 1656, Halley grew up as the child of a wealthy merchant, who educated him at St. Paul's School and Queens College (Oxford). He left Oxford before obtaining a degree but had published his first scientific paper, on the orbits of the planets—in Latin—in the *Philosophical Transactions of the Royal Society.* Oxford did eventually award him an M.A. later that year (1676). Halley continued his scientific career by sailing to St. Helena, a rocky island in the South Atlantic Ocean, using a government grant and transport by the East India Company. Once there, he mapped the constellations of the Southern Hemisphere; this was the first such mapping using a telescope, and upon his return late in 1676, he was elected a fellow of the Royal Society. The astronomical work was published in 1679 as *Catalogus stellarum australium* (Catalog of the southern stars) and the planisphere (model) Halley made of the constellations was presented to the king, Charles II. Recognized as an astronomer, Halley embarked on a tour of Europe, visiting other astronomers, again with support

Edmond Halley (Hulton/Archive/Getty Images)

from influential figures in the court of Charles II. When he returned, Halley married Londoner, Mary Tooke, in 1682.

This polymath made his mark in several fields. One was geophysics, which Halley is credited with founding. One work in that field is "Theory of the Variations of the Magnetical Compass," published in the *Philosophical Transactions of the Royal Society of London* in 1683. Another area of interest to Halley was meteorology. A notable paper on monsoons and trade winds, "An Historical Account of the Trade Winds and Monsoons," appeared in *Philosophical Transactions of the Royal Society of London* in 1686(7). (The date reflects the fact that England had not adopted the Gregorian calendar. Thus, since New Year's Day was March 25 on the old calendar, the

January 1–through–March 24 period is written as 1686–7 or 1686[7]). In his paper, Halley attributed the winds to solar heat, aware that his explanation was incomplete. The development of the first meteorological chart showing wind direction was part of this prodigious output, as were charts of magnetic variation, tides, and the mathematical equations explaining the variation in barometric pressure as a function of altitude.

Interested in all aspects of the physical Earth, Halley studied the rates of evaporation of water and the salinity of lakes and rivers, drawing conclusions from this work about the age of the Earth. A work, on magnetism, "An Account of the Cause of the Change in the Variation of the Magnetical Needle with an Hypothesis of the Structure of the Internal Parts of the Earth," was published in *Philosophical Transactions of the Royal Society*. This work was published in parts between 1687 and 1694, and Halley emphatically disagreed with the ecclesiastical age of Earth, a position that had a serious impact on his career: in 1691, he did not receive appointment to the Savillian Chair of Geometry at Oxford. Halley's career path included several oddities; for a time, he was deputy controller of the mint at Chester (1696–98). The position was one by which the Crown essentially supported his scientific work.

As a fellow of the Royal Society, Halley was the editor of *Philosophical Transactions* from 1685–93, and his work was vital for the continuation of the journal. He was personally involved with the writers of articles and the printers. Also, as editor, he was instrumental in persuading Newton to publish the *Principia* and helped pay the production costs for it. Newton was intensely sensitive. Some criticism of this pivotal work on gravitation and celestial mechanics caused him to refuse to print it. Halley volunteered support for Newton and encouraged him to continue. In addition to his other scientific work, Halley developed actuarial tables, estimated the acreage of England, and the size of an atom—in his day a hypothetical concept. In Halley's time, an atom was the minutest essence of something, anything. He also wrote several critiques of classical authors, detailed descriptions of the aurora, which he called an "effluvium," and surveyed the English Channel.

A scientific vessel, the *Paramore* (or *Paramour*) was built for Halley's use as a floating scientific station. Commanding the ship from 1698 to 1700, he sailed between 52° N and 52° S, mapping magnetic variations. His contemporaries called the connecting lines Halleyan lines; they are now referred to as isogonic lines and are the magnetic analog of isobars, the lines drawn between places of identical barometric pressure. All the points on an isogonic line have the same magnetic inclination—they are in a line pointing toward the north magnetic pole.

Best remembered as an astronomer, Halley published the *Astronomiae cometicae synopsis* (A synopsis of the astronomy of comets) in 1705. This is the work that established the basis for comet study and identified the comet (now called Halley's comet) as one that appeared in 1531, and again in 1607 and 1682. Halley accurately predicted its return in 1758.

Halley finally was appointed to the Savillian Chair in 1703. He also succeeded his rival, John Flamsteed, as astronomer royal. This was Halley's title until he died. The post was an official government appointment, and various duties (and a large stipend) went with it, notably the job of being the official arbiter of latitude and longitude of every place charted.

While serving as astronomer royal, Halley continued his stellar observations and charted over 1,400 lunar meridional transits and a full 18-year lunar cycle. At Halley's insistence, studies of the transits of Mercury and Venus became important to the Royal Society. Their success, in turn, led to the voyages of JAMES COOK and other explorers. The publication in 1731 of *A Proposal for Finding Longitude at Sea . . .* in *Philosophical Transactions of the Royal Society* was motivation for George Graham, an instrument maker, to

lend money to John Harrison for the express purpose of building a chronometer that could be used to determine longitude at sea. The resulting chronometer traveled to the Pacific with Cook. Halley was a pivotal figure in a number of scientific disciplines that deal with the seas. He died in Greenwich on January 14, 1742.

⊠ **Heezen, Bruce Charles**
(1924–1977)
American
Geologist, Oceanographer

Bruce Heezen, the only child of a farming family, was born on April 11, 1924, in Vinton, Iowa. He attended local schools and the University of Iowa, majoring in geology. A fossil-collecting summer project had almost convinced him to major in paleontology when he attended a Sigma Xi (an honorary scientific society) lecture given by WILLIAM MAURICE EWING. Heezen spoke to Ewing after the lecture and was invited to spend a summer on the Mid-Atlantic Ridge. Instantly smitten with the sea, Heezen spent the rest of his life working with Ewing. After earning his degree from Iowa, he went to Columbia University, where he earned a Ph.D. in geology in 1957 and remained in the geology department. His first appointment was as a research associate (1956 to 1958) at the Lamont Geological Observatory. By 1960, he was an assistant professor, then rose to associate professor in 1964. His duties included teaching, but his major interest was research, spending the summers at sea and the rest of the year analyzing the data he had collected. The first major project Heezen worked on was the study of the northeast coast of North America. An undersea earthquake in 1929 with an epicenter around the Grand Banks had severed telephone cables, and he examined the causes for this while he mapped the turbidity currents (see ARTHUR HOLMES) and the patterns in which these currents deposit sand and gravel on the

Bruce Heezen *(United States Navy)*

seafloor. Some of these deposits are very far away from the continental landmass. However, Heezen realized that this was only a partial answer to the question of why there was an earthquake off Newfoundland. The earthquake was a major disaster for the communications industry. Until it happened, the idea that undersea earthquakes occurred and that this made the cables vulnerable had just never arisen.

Sounding studies of the Mid-Atlantic Ridge had begun shortly after the end of World War II. While he was still a graduate student, Heezen and MARIE THARP, a research assistant in the department, began their topographic mapping of the oceans in 1952. This project continued for the rest of Heezen's life. As Tharp and Heezen studied the cable-snapping earthquake, Tharp mapped the rift valley that ran through all of the globe-encircling mountain chain. The valley is one giant, continuous earthquake zone that is now acknowledged as the boundary between crustal plates. This discovery was presented at the 1956 meeting of the American Geophysical Union. It was sensational news. Although fracture zones were known in the Pacific—MATTHEW MAURY had first reported on them—Heezen and Tharp found them in the Atlantic in 1961. These breaks in the undersea mountain range are evidence of undersea seismic activity.

Through the 1960s, Heezen developed theories of new crustal material emerging from the rift valleys; most have passed the test of further observations and correlated with data on seafloor spreading produced by HARRY HESS. An accident in which a coring rig was lost on a 1965 expedition caused Heezen to redesign the explorations and use a camera along with a corer. The camera was specially constructed to withstand great pressure. The use of this equipment made visible the effects of deep currents, the turbidity currents, that move and shape the sediment deposits—the continental drifts—that line the margins of continents. This method was a success, and Heezen went on to use it to map and photograph the drifts in both the Atlantic and Pacific Oceans.

The map produced by Heezen, Tharp, and Heinrich Beran, an artist, was published by the National Geographic Society. It was not just a scientific treasure but a work of art. This publication was followed by *The Face of the Deep* in 1971 coauthored by Heezen with Charles D. Hollister. This book is a classic in descriptive geology. Heezen is responsible for hundreds of photographs of the ocean bottom taken with cameras mounted on ships and from manned and unmanned submersibles. The photographs are important because the pictures are positive proof of the life forms on the ocean bottom. Before this, an argument could be made that in dredging, the dredge could pick up organisms from other layers, contaminating the sample. Also, the dredge destroys what it picks up; it hauls the sample out of context.

Heezen had served as a consultant for the U.S. Navy. He was a member of the Committee for Marine Geology of the International Union of Geological Science, and of the Committee for Marine Geophysics of the International Association of the Physical Sciences of the Ocean. He served a term as president of both of these bodies. He was also a coordinator of the International Union of Geology and Geophysics Institute for its Oceanic Atlas, as well as the recipient of many medals. The navy's interest in his work on deep-sea submersibles led to his association with the navy project that was being conducted near Iceland on the Reykjanes Ridge in the North Atlantic. While working there, Heezen died of a heart attack on June 21, 1977. In 1999, the U.S. Navy launched a military survey ship named for Bruce Heezen.

⊠ **Helland-Hansen, Bjørn**
(1877–1957)
Norwegian
Physical Oceanographer

The son of Kristofer Hansen and Nikoline Helland, Bjørn Helland-Hansen was born in Christiania (now Oslo), Norway, on October 16, 1877. His father was a secondary-school teacher and also a professor at the Royal Frederick University. The university changed its name as it grew, first to the University of Christiania, then to the University of Oslo. Helland-Hansen attended the university in the Faculty of Law and then Medicine. In 1898, while still an undergraduate, he traveled to the north of Norway to observe the aurora. The aurora borealis is a light phenomenon in the night sky of the Northern Hemisphere of the Earth. The Southern Hemisphere's analog is the aurora australis. Both display brilliant colors in the upper atmosphere. The intensity of the displays is in part dependent on the 11-year sunspot cycle and the degree of latitude on Earth.

After Helland-Hansen suffered frostbite, the resulting finger damage ruled out a medical career, and he redirected himself to oceanography. In 1899, encouraged by Johan Hjort of the Norwegian Board of Sea Fisheries, Helland-Hansen went to Copenhagen to study physics with Martin Knudsen, but he never finished the degree; apparently he lost interest in physics. The next year, FRIDTJOF NANSEN was visiting the university and invited Helland-Hansen on an exploratory cruise as an assistant. They worked well together and maintained a lifelong association. From 1900 to 1906, Helland-Hansen was employed by the fisheries board and sailed with Nansen on the *Michael Sars*, taking measurements of the salinity and temperature of the

Bjørn Helland-Hansen *(Photo by Eugene LaFond, SIO Archives/UCSD)*

layers of water. His book about those findings, *The Norwegian Sea*, with Nansen as coauthor, was published in 1909.

Moving layers of fluids, liquids, or gases interested Helland-Hansen. He worked with VILHELM BJERKNES, studying the circulation of fluids of varying densities, and with VAGN EKMAN on wind-driven currents. With Johan Sandström, he created equations to describe ocean currents based on the densities of the water, and the Coriolis force. In 1914, he was part of the design team that planned the research vessel *Armauer Hansen,* and for the next 40 years he used it collecting biological and water samples, mapping currents, and measuring velocity. A principal instrument used was the Ekman current meter, to determine current in the North Sea and the Canary Islands. The careful study of temperature changes in the sea and the effects of the Gulf Stream on North Atlantic waters was Helland-Hansen's principal interest.

Helland-Hansen was director of the biology section of the Bergen Museum from 1914 to 1946, and professor of oceanography at the museum, which eventually became the University of Bergen. He was director of the Geophysical Institute at Bergen from 1917 to 1947, and was active in the International Union for Geodesy and Geophysics, serving as president from 1945 to 1948, and continuing throughout his career publishing of articles on oceanic temperatures, contents, and layers. He was a directing member of the International Council for the Exploration of the Sea from 1947 until he died in Bergen on September 7, 1957.

⊠ **Hensen, Viktor**
(1835–1924)
German
Physiologist, Oceanographer

When Hensen was born in Schleswig on February 10, 1835, it was a part of Denmark. Young Viktor was educated locally, and began his medical studies in Würzburg, moved to Berlin's medical school for a year, and then to Kiel. He graduated from Kiel in 1858, receiving a physiology degree, and remained in the physiology department there. He attained a professorship in 1871 and kept that position until he retired in 1911.

As a physiologist, he worked on the physiology of sensation. One paper in 1862 described the hearing organs in decapods (lobsters, shrimp, and crabs), and another in 1865 described the eyes of cephalopods (squid, and octopi). Another area of his physiological research led to his discovery of glycogen, the compound manufactured and stored in the liver of vertebrates. It is an instant energy source.

However, he is best known for his research in marine biology; his interest in the field dated from 1863. Once intrigued by the multitude and physiological complexity of the organisms, he quickly became a self-taught professional.

In addition to straightforward descriptions of organisms, Hensen developed quantitative methods for assessing the commercial fish populations, developing the Hensen net, fine mesh filter that could capture the unicellular organisms in a square meter of water. Since plankton are so small, they would pass through an ordinary cloth net. This, again, was a quantitative sampling technique. The equipment was tested in the North Sea and the Baltic in 1885. Inaugurating the term "plankton" to describe the tiny marine organisms, both plant and animal, in 1887, Hensen used his equipment to sample ocean waters and in 1889 worked extensively in the Atlantic. One of his more intriguing results was the discovery of greater concentrations of plankton in temperate rather than in tropical water. Earlier logic would have argued for the reverse situation.

All of Hensen's publications were in German. Those directly involved with marine science include *Ergebnisse der in dem Atlantischen Ocean von mittle Juli bis enfang November 1889 ausgefürten Plankton Expedition der Humboldt*

Stiftung (Result of the Humboldt Institution's Atlantic Ocean plankton expedition, from mid-July to the beginning of November, 1889), 1892; "Betreffend der Fischfang auf der Expedition" (Findings on fisheries) in *Jahresbericht der Kommission zur wissenschaftlichen Untersuchung der deutschen Meere im Jahre 1871* (Annual report of the commission on scientific research in the North Sea in 1871), 1873. Data from a later cruise yielded "Das Plankton der Östlichen Ostee und des Stettiner Haffs im 1887–91" (On plankton of the eastern Baltic Sea and the Stettin Bay in 1887–91), published in 1893 in the Annual Report of the Commission on Scientific Research in the North Sea. *Über die quantitative Bestimmung der Klieneren Plankton-organismen in wissenshaftliche Meersuntersuchungen* (On quantitative determinations of the smallest plankton organisms in scientific oceanography), published in 1901, and his collected methodology, *Die Methodik der Plankton Untersuchung in Handbuch der biochemischen Arbeit-Methoden* (The methodology of plankton research in the handbook of biochemical procedures), was published in Vienna in 1911.

Recognition of his quantitative studies brought Hensen a fellowship in the Leopoldine Academy of Berlin, and the Prussian and Bavarian Academies of Science. He was also a member of the Faculty of Medicine at Kiel. Remaining at Kiel after his retirement, Hensen died there on April 5, 1924.

Harry Hess *(Princeton University Library)*

⊠ Hess, Harry Hammond
(1906–1969)
American
Geologist

Harry Hess was born in New York on May 24, 1906, and raised in Asbury Park, New Jersey. He began his education at Yale in the Electrical Engineering Department, later moved to geology, graduating with a B.S. in 1927. After about two years as a field geologist in northern Rhodesia

(now Zambia), Hess returned to the United States for graduate study, first at Rutgers and then at Princeton, where he worked with FELIX VENING MEINESZ, and was awarded a Ph.D. in 1932. After a brief period in Washington, D.C., at the Carnegie Institute, doing geophysical research, Hess returned to Princeton as a faculty member in the Geology Department, a position he held for the rest of his life. Hess was chairman of that Geology Department from 1950 to 1966. Navy service in World War II led to his continuing in the navy as a reserve officer: Hess rose to the rank of rear admiral in the navy reserves.

While serving in the wartime navy, Hess began echo soundings in the Pacific, bouncing a sonar wave of known frequency straight down to the ocean bottom and listening (using acoustical equipment) for the wave bouncing back. Since sound travels at a known rate in water, one can calculate the distance to the bottom by the time it takes for the sound wave to return to the source. It was then that he discovered the mesa-like, flat-topped undersea elevations that had no coral reefs on them. These hitherto unknown seamounts Hess named guyots, in honor of the first professor of geology at Princeton, Arnold Guyot. Explaining the guyots began an evolution of Hess's thinking about their creation and continued existence, in conjunction with the fact that there was relatively little sediment on the ocean floor, and no really old fossils—nothing older than the Cretaceous—while marine fossils much older than that have been found on land. In 1962, Hess published a pivotal paper, "The History of Ocean Basins," wherein he presented the theory that the ocean floor is young and constantly being created by rising magma. More work developed this theory, showing that it further explained the spread of continents and the relative lack of sediment on the sea bottom. This work also continued the efforts to validate ALFRED WEGENER's theory and the work of ARTHUR HOLMES, who proposed a theory of convection currents as an explanatory mechanism. Wegener advanced the idea that the floor of the ocean was the newest part of the Earth: new edges of plates were being formed and then moved. This explained the similar fossils on now widely separated continents. Holmes kept Wegener's theory current by suggesting a mechanism for plate movement. Hess built on the research of these earlier geologists.

Hess received many awards and recognition for his contributions to geology. He was elected a fellow of the National Academy of Science, the American Philosophical Society, the Geological Society of London, and a medalist of the Geological Society of America. He was president of the American Geophysical Union's Section of Geodesy and Technophysics (1951–53) and

of the Mineralogical Society of America. Urged on by WALTER MUNK, Hess became an enthusiast for the Mohole Project and worked to have it funded by the National Science Foundation from 1958 through 1966. The Mohole project was an international effort to drill down to the Moho layer. This discontinuity in the body of the Earth separates the crust from the underlying mantle; it was discovered by ANDRIJA MOHOROVIČIĆ. Hess held another post, as chair of the Space Science Board of the National Academy of Science. As an adviser for NASA, he was attending a meeting of the Space Science Board in Woods Hole, Massachusetts, when he suffered a fatal heart attack on August 25, 1969.

⊠ **Holmes, Arthur**
(1890–1965)
British
Geologist

Physics was the first scientific subject that interested Arthur Holmes. Born in Hebburn-on-Tyne, a small farming village, on January 14, 1890, he studied at the Imperial College, London, graduating in 1910. While an undergraduate, he changed his major from physics to geology. Upon graduation he went as a graduate student to Lord Rayleigh to study radioactivity. In 1913, just before finishing his doctorate, he proposed the first geologic time scale based on radioactivity. The age of Earth had been a vexing topic long before the work of CHARLES LYELL, the premier English geologist of the 19th century. Using the concept of radioactive decay as one component of the heat of the Earth's interior, Holmes arrived at an estimate of the age of the Earth that was far greater than that suggested by any other geologist. The estimate of Earth's lifespan was then about 16 million years, but Holmes's is still the figure used: about 4 billion years. The only alteration is in the beginning of the Cambrian period; Holmes had placed that at 600 million years ago, but it is now thought to be slightly more recent, 590 million years ago, due to one of the several revisions he made in the time scale.

Holmes took a scientific journey to Mozambique in 1911, followed by teaching at the Imperial College until 1920, when he tried an industrial position and went to Burma as an oil geologist. On his return in 1925, he became a professor of geology at the university of Durham. In 1943, he moved to Edinburgh, Scotland, where he was also professor of geology. He held both positions for the rest of his life.

In 1929 Holmes reinvestigated ALFRED WEGENER's theory of continental drift and believed that he had a mechanism by which the crustal plates of Earth could be moved. Holmes based his idea again on radioactivity, which would heat the mantle, moving it upward by convection currents and thus move the plates. When the mantle plume cooled, Holmes hypothesized, it sank again into the interior. This was only theory. It would take the work of ROBERT DIETZ, HARRY HESS, MAURICE EWING, BRUCE HEEZEN, DRUMMOND MATTHEWS, and TUZO WILSON to substantiate this view of the Earth's makeup.

Holmes's greatest contribution to science was his serious consideration of Wegener's theories. Wegener was not taken seriously by geologists because he was a meteorologist. Holmes established the theory of crustal plates as a significant geological idea. His major publication was the book *Principles of Physical Geology*. First published in 1944, it was thoroughly revised and reissued before his death in Edinburgh on September 20, 1965.

Humboldt, Friedrich Wilhelm Karl Heinrich Alexander, baron von
(1769–1859)
German
Explorer, Natural Scientist

Born in Berlin on September 14, 1769, Alexander von Humboldt was an indifferent student des-

tined for an engineering education when he became entranced by botany and spent a year at the university in Göttingen. Natural science was then a "hot" field. By contrast with the new and exciting finds that were being made almost daily, engineering was dull.

However, he continued his studies in engineering, graduated, and worked in that field, until an inheritance in 1796 made it possible for him to retire. After an intensive preparatory program of study, and in spite of the difficulties in obtaining materials and transport in the midst of the Napoleonic wars, Humboldt sailed for South America in 1799. His botanist companion on this five-year-long journey was Aimé Bonpland, and together they traveled more than 6,000 miles through what are now Venezuela, Peru, Ecuador, and most of Central America and Mexico. The very long trip to South America was the scientific opening of the continent. Before his explorations, the coast had been charted only minimally, and little was known of the interior. The conquering nations, Portugal and Spain, had done almost nothing other than use the territory as a source of raw materials.

Settling in Paris after his return in 1804, Humboldt spent the next 20 years writing 30 volumes based on his voyage. Humboldt's writings were in large part a series of diaries of his voyages, entitled *Voyages aux régions équinoctiales du nouveau continent fait en 1799–1804* (Voyages made to the equatorial regions of the new continent in 1799–1804). This series of books began to appear in 1807 and continued until 1833. Some 12,000 pages in 30 volumes were completed, and the publication cost exhausted Humboldt's funds but established him as a scientific phenomenon. He was truly a member of a small charmed circle of scientific superstars who were recognized as such worldwide. People came from other countries to visit them. Humboldt, GEORGES CUVIER, and Joseph Gay-Lussac, a chemist, were such recognized scientists. Humboldt was honored by every scientific society in Europe and America.

So many Americans visited him that he joked he was becoming half-American.

After his money was depleted, he returned to Berlin as tutor to the crown prince in 1827. This position led to his next commission in 1829, a tour of several months' duration to explore the resources in Russia and Siberia. CHRISTIAN EHRENBERG accompanied him as one of his assistants. The trip to Siberia was commissioned by the Russian government. Siberia was a vast region largely unknown to the European Russians, and the government wanted to know what resources were available in this vast, freshly subdued region. Upon his return to Berlin eight months later, Humboldt began writing *Kosmos*—a compendium of all scientific knowledge. Four volumes were published by the time of his death; a fifth volume was published posthumously. After his move to Berlin, he was again in the center of a scientific world, at the Leopoldine Academy, the University of Berlin, and the Prussian court. *Kosmos* became the standard work on the integration of all sciences and the placement of man. In this work, man is a part of the natural order of the universe. Humboldt's writings were popular literature. They were translated into many languages; in some translations, the author is given as Alexander Humboldt. Some of these writings are still in print.

One of Humboldt's many attributes was that he wrote in a very entertaining and engaging style. He could interest his readers through his descriptions of exotic locales while instructing them in real science. Among his accomplishments were the description of the cold current on the western coast of South America, then called the Humboldt (now the Peru) current; the proposition and geologic assessment of the feasibility of a canal across the Panamanian isthmus, and the relationship of volcanic activity and mountain building. Humboldt's definitive idea ended the geological theory of neptunism; until his work, neptunism had been a viable theory explaining the geologic history of Earth. In the neptunists'

view, the Earth had once been totally covered with water that has since receded.

In Humboldt's meteorological work, he noted the inverse relationship of altitude and temperature, and its effect on the local flora. Thus began both the research linking particular places with their specific biomes, and significant weather studies. His climate maps were the first to include isotherms and isobars as vital bits of information that explained weather. Baron von Humboldt died in Berlin on May 6, 1859.

⊠   Hunt, Maurice Neville
(1919–1966)
English
Geophysicist

Cambridge, England, was Maurice Hunt's home. He was born there on May 29, 1919, and lived there all of his life; his parents were Archibald Hunt, the professor of physiology at the university, and Margaret, his mother, the sister of John Maynard Keynes—the world-famous economist. Hunt was educated in the Highgate schools after the family moved to London and spent summers with his father at the Marine Biological Association's laboratory. Before starting at Cambridge in 1938 for a degree in natural science, Hunt sailed to Bermuda, where he conducted weather and water studies on the Royal Society's research vessel *Culver*. When World War II began in 1939, he was in the Royal Navy in a technical post. Hunt resumed his academic career in 1945 when he returned to Cambridge for a year of physics and a wartime (accelerated) B.A. in 1946. The Ph.D. program he then began was in geodesy and geophysics, where his task was to develop a methodology for deep-sea studies of sediment thickness and underlying bedrock. Geodesy is the study of the shape of the Earth and exact measurements of points on its surface.

In 1949 he had been elected a fellow of Kings College and research assistant in the department of geodesy and geophysics. Modeling his work on W. MAURICE EWING's two-ship method, Hunt used one ship in shallow water and two for deeper soundings down to the Moho layer, completing this work in 1952, a year after finishing the thesis for his Ph.D. The aim of these soundings was to establish the depth of the Moho layer. It is closer to the Earth's surface on the ocean floor than it is on continents. While engaged in Moho studies, he was also involved in the work of *Challenger II* (1950–53). This voyage, like its predecessor, was commissioned by the British Admiralty, with the charge that it conduct another series of scientific inquiries of the ocean. The *Challenger II* expedition was a series of seismic studies conducted undersea worldwide, and Hunt was personally involved in those of the northeast Atlantic.

Throughout the 1950s, he continued to study sediment cores and brought bedrock samples to the surface, thus providing the raw material for the work of FREDERICK VINE and DRUMMOND MATTHEWS on geomagnetism. Drills

were lowered to the ocean bottom, and rock samples were drilled out and suctioned to the surface. From these samples, Matthews and Vine deduced the changing magnetic lines of force. The rock sampling was an integral part of the program of the International Indian Ocean Expedition (IIOE) and involved the participation of a large group of Cambridge scientists. Most of this work was presented at a Royal Society meeting arranged by Hunt in 1962.

That same year, Hunt was elected a fellow of the Royal Society and was the 1963 medalist of the Physical Society. Advancing to a reader's position (the American equivalent is associate professor) in marine geology in 1965, Hunt was in the midst of a brilliant career when profound depression led to his suicide on January 11, 1966. His writings were incorporated in the publications of the IIOE. He had contributed to the understanding of the nature of the sea bottom and provided the essential evidence for the breakthrough work on magnetic reversals that confirmed the theory of plate tectonics.

J

⊠ Jeffreys, John Gwyn
(1809–1885)
British
Zoologist

Born in Swansea, Wales, on January 18, 1809, John Jeffreys followed his father to establish a career in the law. While pursuing biological studies as a hobby, he was a solicitor, and in 1856, became a barrister in Lincoln's Inn, London. Jeffreys continued his law profession until his retirement in 1866 when he decided to devote all his working time to biology.

Jeffreys's reputation in biology began when he was 19, with a work on British mollusks presented to the Linnaean Society. At the time, 1829, most scientific information was presented as talks at meetings of learned societies. The creation of the periodical literature came later. The next year, 1840, he married Ann Nevill and was elected a fellow of the Royal Society. His interest in mollusks continued, and Jeffreys traced the modern forms of this phylum to their ancestors in the Tertiary period, producing many papers and a five-volume work entitled *British Conchology* (1843).

While EDWARD FORBES's idea that life ended at a depth of about 300 fathoms (600 m or 1,800 ft) was collapsing in the 1860s, Jeffreys in 1868 embarked on a collecting cruise on the HMS *Lightning*. The living specimens collected were hauled up from depths greater than 600 fathoms. This was definitive proof that Forbes was wrong about the absence of animal life at great depths. Since Jeffreys found living organisms at depths of 600 fathoms, he became more certain that this limit was an error; he embarked on a program of sampling at a variety of depths. The water sampling also disproved the idea then current that all seawater at great depths was at a constant temperature of 4° C.

Other voyages with Jeffreys in charge were conducted in 1869 and 1870 on the British Admiralty's vessel, HMS *Porcupine*. The amassed specimens of a number of different phyla came from a succession of depths as far down as 2,400 fathoms (4,364 m or 14,400 ft). The water samples from all depths presented a range of temperatures—confirming his earlier work. These cruises were restricted to British waters and yielded more new species than had been found in the same area by the HMS *Challenger* expedition of 1872–76 and provided Jeffreys with raw material to study for the rest of his life. Other collecting trips took Jeffreys in 1875 to the Arctic on the HMS *Valorous* and in 1880 on the *Travailleur*, a French vessel, to the Bay of Biscay. The numbers of new species brought Jeffreys to the conclusion that there are more mollusks than anyone dreamed of, and they varied significantly with locality and environmental conditions.

Jeffreys was a founding member of the Marine Biological Association of the United Kingdom in 1884. He continued to study the Mollusca and intended to donate his splendid collection of European mollusks and their fossils to the British Museum. However, the collection was sold and is now in the Smithsonian Institution in Washington, D.C. Jeffreys died in London on January 24, 1885.

⊠   **Johnson, Douglas Wilson**
(1878–1944)
American
Geomorphologist

Douglas Wilson Johnson was born in Parkersburg, West Virginia, on November 30, 1878. His parents were Isaac Hollenback, a lawyer, and Jane Amanda Wilson Johnson, a strong supporter of the temperance movement. Orphaned at 12, he was able to attend Denison University on a scholarship. However, after Johnson's freshman year, Clarence Derrick, a geologist who had been at Denison, invited Johnson to spend a summer assisting him in geological sampling in New Mexico. Johnson transferred to the geology department at the University of New Mexico and after graduation continued his education with a graduate fellowship. He was awarded a Ph.D. from Columbia University in 1903. First appointed to the faculty at the Massachusetts Institute of Technology, directly after leaving Columbia, Johnson worked with W. M. Davis of Harvard on physical geography, edited Davis's work, and then transferred to Harvard in 1907. Physical geography is the description of landforms: its rainfall, its elevations and the composition of its rocks, where the mountains are and their height, or what is the average annual rainfall. The major interest then was shorelines and beaches: how they form, change, and move. This work continued thanks to a Shaler grant. Another appointment brought Johnson back to Columbia as a faculty member in 1911, and he was full professor there in 1919 when he published a major book, *Shore Processes and Shorelines*. This work was followed in 1925 by *New England—Acadian Shoreline*. The two books were well received. A number of significant papers followed between 1910 and 1930, and the importance of these works in geographic research led to the formation of the National Research Council's Committee on Shoreline Investigations, with Johnson as chair. The goal of the committee was to research sea levels and shoreline changes.

Another major publication was *Principles of Marine Level Correlation* (1932) on the construction and conformation of submarine terraces. A continental shelf, such as the one off the Atlantic coast of North America, is characterized by a broad, gently downward slope of bottom. This may be marked by one or more terraces, and a level stretch of bottom, followed by a steep downward slope, followed by another fairly level stretch of ocean bottom or plateau before the steep slope to the bottom. The analog on land is mountain terracing. Johnson went on to studies of the influence of topology on warfare. World War I left the French countryside devastated. It was not just buildings that were destroyed; riverbeds were blown up and their drainage totally altered, hills were leveled, and lakes had outlets clogged with debris that turned them into swamps. Johnson was part of the effort to repair the ruins that had been created in the landscape by the bombardment. The French government acknowledged his efforts with the Legion of Honor's Medal of Valor.

Later, he was president of the Commission on Terraces of the International Geographical Union, of the Geological Society of America, and of the Association of American Geographers. He was working on the Florida coastline when he died of a heart attack in Sebring, on February 20, 1944.

K

Knipovitch, Nikolai Mikhailovitch
(1862–1939)
Russian
Hydrologist, Marine Biologist

Nikolai Knipovitch was born in Suomenlinna on March 25, 1862. His birthplace is part of that region that was contested by several ruling powers: it had been controlled by Russia at times, by Sweden at times, and is now in Finland. The future scientist became a student at the Faculty of Science in the University of St. Petersburg; he graduated in 1886. In the year before his graduation, he was a member of an 1885 expedition to the Grimm area on the Volga River. This research survey explored the commercial possibilities of the herring fishing industry. The voyage continued down into the Caspian Sea and was the beginning of Knipovitch's continuing interest in hydrology and the marine life it supported.

In 1887, Knipovitch was assigned to a White Sea expedition. The area of study was enlarged to include the adjoining Barents Sea, and its aim was again to determine the size of the edible fish population in this arm of the Arctic Ocean. This limited goal of the government was to be the constant theme in Knipovitch's work and remained so even after the revolution in 1917. The government sponsorship of exploratory cruises extended only to the determination of the resource values of a particular area. This is not an aim of science; it is at best the application of scientific information to commerce. Knipovitch managed to do scientific work in spite of the limited goals of the expeditions he conducted and the meager budgets allotted. While on the 1887 assignment to the White and Barents Seas, he developed his methodology for studying the interaction of the water of an area, its temperature, currents, dissolved minerals, and ice level, as well as the biota, the plant and animal populations that lived in it.

Continuing his total organism studies, Knipovitch spent most of the years from 1898 through 1901 aboard the *Andrey Perozvanny*, a ship built for him, which was reinforced to withstand the pressure of ice freezing around it. The officially stated aim of these expeditions to Murmansk, a deepwater harbor in the far north of European Russia, was to assess the local waters as a food resource. This again was a fish-counting expedition which its leader, Knipovitch, turned into a study of the food web of the region.

Gaining recognition from his work in the Arctic, Knipovitch was assigned to study the fish-producing waters in more temperate regions. He organized one trip to assess commercial possibilities of the Caspian Sea in 1904 and was then shifted to the Baltic. There he did two all-populations studies, one in 1905 and another in 1911. (An all-population study looks at everything in a biome and attempts to understand all the interactions between all species of plants and animals and the earth and water they live on or in.) He was redirected to the Caspian Sea, and worked there collecting specimens and information in the 1912–13 winter, and again the next year, 1914–15. He returned to St. Petersburg to write his report of these voyages.

After the change of government in 1917, Knipovitch managed to maintain his position at the Faculty of Sciences in St. Petersburg, renamed Petrograd. His last voyages were again to the southern seas, several to the Sea of Azov and the Black Sea between 1922 and 1927, and a last trip to the Caspian in 1931–32. The purpose of all of these trips was the assessment of their exploitable fish to feed a global market.

In spite of the limitations of his funding, Knipovitch managed to publish 164 works, all in Russian. He was the first Russian scientist to correlate the study of marine fish with the hydrology of the waters they lived in. His detailed studies of what the fish ate and what fed on them were an excellent contribution to the literature of the interdependence of organisms and their relationship to the waters that sustain them. Knipovitch died suddenly on February 23, 1939, in Leningrad, which has now returned to its original name, St. Petersburg. Translated titles of some of his books include *Principles of Hydrology in the European Arctic Ocean*, St. Petersburg, 1906; *Hydrological Explorations in the Caspian Sea*, St. Petersburg, 1914; *Transactions of the Caspian Expedition . . .*, Petrograd, 1921; *Checklist of the Fishes in the Black and Azov Seas*, Moscow, 1923; *Checklist of Fishes in the Barents, White and Kara Seas*, Moscow, 1926; *Hydrological Explorations in the Black Sea*, Moscow, 1933; and *Hydrology of the Seas and Saline Waters . . .*, Moscow-Leningrad, 1938.

⊠ **Kofoid, Charles Atwood**
(1865–1947)
American
Biologist

The son of Nelson and Janet Blake Kofoid, Charles was born in Granville, Illinois, on October 11, 1865, and spent most of his adult life on coastlines. He attended Oberlin College and, after graduation in 1890, went to Harvard for a Ph.D. in biology. After a year at the University of Michigan, Kofoid became director of the biological station of the University of Illinois and studied river plankton. By 1901, he accepted a position in the zoology department of the University of California and became its chair in 1910. He held that post until 1936, when he retired. His only time away from the department were the years 1919 through 1923, when he was on leave in Japan, studying the Japanese species of boring worms.

While he was a young associate professor, Kofoid had accompanied ALEXANDER AGASSIZ on a collecting trip in the Pacific. Kofoid was one of the scientific crew on the 1904–05 cruise of the USS *Albatross*, the research vessel whose scientific commander was Agassiz. The trip was designed to find, collect, and classify organisms. In doing this, the crew discovered similar but not identical organisms in regions isolated from each other. This reinforced the work of CHARLES DARWIN, who had expounded the theory of common ancestry. Similar organisms, close to each other but separated by some barrier, showed changes in some characteristics, and those changes could best be explained as a change in inherited characteristics that was locally advantageous. Evolution and genetic drift—gradual change in morphology because of small changes (mutations) that make an organism different from its original ancestor—still had to be proved to many people, including biologists. Kofoid and Agassiz's work was done only 45 years after Darwin's epochal book on evolution and before the redis-

Charles Kofoid *(SIO Archives/UCSD)*

covery of Mendel's works explaining the mechanism of inheritance.

While sailing on the *Albatross*, Kofoid redesigned collecting equipment, perfecting a horizontal net for retrieving a sample at a uniform depth, as well as a plankton-collecting bucket that takes a vertical sample of a column of organisms, each living at a particular depth, at a particular time of day. Plankton migrate in relation to available light.

Another long trip was conducted in the winter of 1915–16 when Kofoid went to the Far East to study marine parasites. This was followed by a study of the shipworms in San Francisco Bay. This work was of great commercial interest; shipworms are a major problem wherever there are wooden pilings, docks, and boat hulls.

A collector of books on marine biology, invertebrate zoology, and parasitology, Kofoid do-

nated his important collection to the biology department of the University of California at Berkeley as a nucleus of a distinctive library. His area of expertise was plankton and other pelagic organisms—those found in open ocean—of the Pacific. He was a teacher and mentor to more than 60 graduate students. The editorship of the University of California's Publications in Zoology was Kofoid's permanent post. He was one of the founders of *Biological Abstracts*, the significant index in the field, begun in 1927; he edited the General Biology Section of this publication for many years. He helped to found the Scripps Institution of Oceanography at La Jolla, California—an enduring legacy.

Kofoid's research started with freshwater plankton. Much of this work was collected in large monographs finished long after he moved to California and published as "Plankton Studies, I–V," in the *Bulletin of the Illinois State Laboratory for Natural History*, 1894–1908. In California, his interests moved to marine plankton, particularly the Tintinnida (unicellular protozoans with a bell-shaped test, or shell), and dinoflagellates (flagellated protozoans with ovoid shells). Continuing with analogous marine organisms, he published "Some New Tintinidae from the Plankton of the San Diego Region," in *University of California Publications in Zoology*, 1905; and "Dinoflagellata of the San Diego Region, I–V," in the same series, 1906–08. Another large work on dinoflagellates presented the results of the *Albatross* voyage, and it was published in the *Bulletin of the Museum of Comparative Zoology*, 1911. Much of the work on protozoology was done with Olive Swezy, a long-term coworker.

Another interest, while not marine science, brought Kofoid national recognition. He became involved in work on intestinal parasites, and in 1918 was appointed a major in the U.S. Army's Sanitary Corps, assigned to make a hookworm study and organize a parasitological laboratory. This led to other work for the federal government, reporting on similar laboratories in European stations He reorganized his examination of the life cycles of human parasites, and their relation to public health. Working with parasites, he became quite expert on the subject of marine borers, such as *Teredo*, which caused considerable damage in the California shipyards. His research on the subject, funded by the National Research Council, led to publication of *Marine Borers and Their Relation to Marine Construction of the Pacific Coast . . .* in 1927, published by the National Research Council, with Kofoid and C. L. Hill (another long-term associate) listed as coeditors.

Kofoid's major effort in founding the abstracting and indexing tool *Biological Abstracts* meant that there was a mechanism with which any researcher could find the work of another on a particular subject without paging through years of many journal volumes. Charles Kofoid died in Berkeley on May 30, 1947.

⊠ Kuenen, Philip Henry
(1902–1976)
English/Dutch
Geologist

The third of five children of Johannes Petrus Kuenen and Dora Wicksteed, Kuenen was born on July 22, 1902, in Dundee, Scotland, where his father was a professor of physics. The bilingual, binational family moved to Leiden, Holland, when Philip was a boy. He was educated in Leiden and took his degrees in geology at the university there, the B.S. in 1922 and the Ph.D. in 1925. Greatly influenced by the work of B. G. Escher, an experimental geologist, Kuenen remained in Leiden after his degree to work as an assistant in the Geology Department from 1926 to 1934. During that time, he was away from 1929 to 1930 on the *Snellius* expedition to the Moluccas, an area of spectacular ocean deeps in what was then the Dutch East Indies and is now Indonesia. Kuenen's task on this voyage was to collect data and use the new sounding equip-

ment. In 1932, he married Charlotte Pijzel, who organized Kuenen's life outside of his work and traveled with him everywhere.

Leaving Leiden for an appointment at the university in Groningen, Kuenen was the curator of the geology collection there and also the lecturer for the undergraduates in geology. There were very few students; this gave him more time for research. Advancing to reader (a position equivalent to that of associate professor) in 1939, he finally achieved professorial rank in 1946. This advancement had been delayed for three years; the pro-German administration had tabled it. Kuenen had been an observer for the Dutch armed forces and was wounded in 1940, and it was known that he was not kindly disposed to the German occupation.

The researches Kuenen had started with Escher in 1929 involved the study of salt domes through the 1930s and 1940s. He then continued the work begun on the *Snellius* voyage, examining gravity anomalies and the creation of undersea volcanic cones. The first experiments on surface configurations and examination of how materials are transported started with a study of glaciers. This led Kuenen to work on turbidity currents, beginning in 1936 and continuing for the rest of his scientific working life. Again, the work on board the *Snellius* was formative since it led to study of the growth of coral reefs and the controls exerted on that growth by changes in sea level and temperature. Two popular works and his diary were published after the voyage; *Kruistoch ten over de Indische diepzeibekkens* (Cruise in Indian deepsea basins) in 1941, and *De Kringloop van het water* in 1948. The latter appeared in revised form in English as *The Realm of Water* in 1955. A popular textbook, *Marine Geology*, was published in 1950. The explosion of information about the geology of the sea rendered this quickly outdated, but Kuenen never found time to update it.

Continuous echo sounding, used from the 1930s onward, revealed the complexity of the ocean bottom, and the number and size of the submarine canyons. Kuenen became interested in the study of turbidity currents in the canyons because of remarks made by FELIX VENING MEINESZ about their ability to erode the canyons. Kuenen began a series of experiments on mud-carrying currents. He dug a steep trench in his garden and used a garden hose to create them. Work on this topic and his books occupied him during World War II, and in 1947, he published a major summary of his turbidity work. The full reports of these were presented first at the International Geological Congress in London (1948) and were the sensation of the meeting. Kuenen demonstrated the mechanisms of transport of coarse sediment out of the canyons and onto the deep seafloor. This was the beginning of new thinking about the construction of graded sand beds. C. L. Mighorini, in the audience, was much impressed. Mighorini had worked on graded gravel beds in Italy in 1944, and he invited Kuenen to join him on an exploratory trip to the Apennines. Their joint paper, "Turbidity Currents as a Cause of Graded Bedding," appeared in the *Journal of Geology* in 1950. It was a major success.

Kuenen extended his work in the area of turbidity currents to other parts of the world; a stay in California produced *Sedimentary History of the Ventura Basin* (1951). The findings of W. MAURICE EWING and BRUCE HEEZEN led to Kuenen's explanations of the phenomenon of transatlantic telephone cable breaks. The transatlantic cable was a technological marvel of the 19th century. It made communication, and therefore commerce, between North America and Europe instantaneous. An unexplained break in the cable was a serious matter that needed careful study, not just to repair the break but to understand why it happened and try to forestall future ruptures. On a more popular level, Kuenen wrote the articles on the continental shelf and the continental slope for the 1973 edition of *Encyclopaedia Britannica*.

After his wife died in 1967, this very quiet man became even more so, and his research slowed dramatically. While colleagues and upper-division students found him helpful and supportive, he had no time or patience for any undergraduates. He retired to Leiden after his wife's death, and died during surgery there on December 17, 1976.

Kylin, Johann Harald
(1879–1949)
Swedish
Botanist

The family name of Johann Kylin was really Olsson: the future botanist, the oldest son of 10 children, was born on February 5, 1879. He changed his surname to that of the family's farm. The newly named Kylin graduated from gymnasium (secondary school) in Göteborg, Sweden, in 1898, and went on with his education at the university in Uppsala, where he received a Ph.D. in 1907. After a year as docent (teaching assistant) at Uppsala, he taught at local secondary schools and teachers' colleges until an appointment in the Botany Department brought him to the University of Lund in 1920. Kylin remained in the Botany Department until his retirement in 1944.

Beginning with his thesis on the marine flora of Sweden's coast, Kylin continued this study for his entire career. His definitive work on Rhodophyta, a large and diverse phylum of red algae, was finally published posthumously by his wife, Elsa, also a biologist. Johann Kylin had created a massive reorganization of the green, brown, and red algae of the Swedish coasts, and wrote a series of papers on algal physiology, pigments, acid-base reactions, chemical composition, and osmotic properties. This work led to several collecting trips to the United States—to California and Woods Hole, Massachusetts—and to England and France, to compare results with colleagues. His work led to a vast reorganization of all the algae; they are no longer considered plants but are in phyla all their own. The red algae in particular, those that were Kylin's principal interest, are a diverse group: the red pigments are compounds that attract increasing attention from pharmaceutical chemists because they are both visible signs of plant stress and they react with free radicals (toxic molecular fragments) to render them harmless.

Kylin was awarded honors by the Royal Society, the Academy of Science of the United States, the Royal Danish Academy of Science, and the Botanical Society of America. He was still working on the Rhodophyta when he died in Lund on December 16, 1949.

L

Lacépède, Bernard-Germain-Étienne de la Ville-sur-Illion, comte de
(1756–1825)
French
Natural Scientist

Lacépède was the only child of Jean-Joseph Médard, comte de Lacépède, who had taken that title from a granduncle upon becoming his heir. Taught at home, Bernard showed early talent for both science and music. Pursuing the latter, he attempted a career as a composer of opera, but was disappointed by poor reviews and lack of sponsors. When GEORGE BUFFON invited Lacépède to continue Buffon's book, *Histoire naturelle* (Natural history), Lacépède accepted and was assigned to work on classification of the reptiles and fish. A position as *gardien* (equivalent to curator) at the royal collection in the Jardin des plantes (the botanical garden) was in effect a subsidy that enabled him to proceed with the work on classification.

During the Revolution, Lacépède served in the Chamber of Deputies and was involved in the planning for and reorganization of the Jardin des plantes which became the Musée d'Histoire Naturelle (Museum of Natural History). He was forced out of the planning committee in 1793 and left Paris to escape the Reign of Terror, the most

Bernard Lacépède *(E. F. Smith Collection, Rare Book & Manuscript Library, University of Pennsylvania)*

radical and bloodiest period of the Revolution. Meanwhile, his work on the *Histoire naturelle des poissons* (Natural history of fishes) continued. By 1795, Lacépède had married Anne Gautier, a friend's widow, and returned to Paris, where he assumed the post of chairman of zoology at the museum. He was working in three areas: teaching, substantially increasing the collection and classifying it, and serving in government. The teaching aspect of his work was eventually given to one of his assistants in 1803, when Lacépède returned to finishing Buffon's work. The five volumes on the fish appeared between 1798 and 1803. This was followed quickly by the work on *Cétaces* (Whales) in 1804. At the time, whales were classified with fish because both swim.

Government service took up a major part of Lacépède's life. He served as senator from 1799, retiring to Epinay-sur-Seine with the fall of Napoléon in 1815. With the return of the monarchy, Lacépède, *comte* once again, was a peer of France serving the new government and wearing the medals awarded him as grand chancellor of the Legion of Honor in 1803. He died at Epinay on October 6, 1825.

⊠ Lamarck, Jean-Baptiste-Pierre-Antoine de Monet, chevalier de
(1744–1829)
French
Natural Scientist

Born on August 1, 1744, in Bazentin, Jean-Baptiste-Pierre-Antoine de Monet Lamarck was the youngest of the 11 children of Philippe-Jacques and Marie Françoise. The family was part of the impoverished lesser nobility of Picardy that had a strong military tradition. Lamarck's father and several of his brothers were in the army. The boy was sent to the Jesuits' seminary in Amiens in 1756, but after his father died, he ran away and joined the army. He was 15 but already an experienced collector of botanical specimens. As the army then moved him from one place to another, his collection grew, and he correlated the differences in flora with region.

Leaving the army in 1768, Lamarck worked in a Paris bank and studied botany and medicine.

At that period, studying medicine involved attending occasional lectures and anatomical demonstrations. He continued his botanical studies and finally finished writing a comprehensive *Flore françoise* (Plants of France) in 1779. The plants were classified by a non-Linnaean system created by the Jussieu family of botanists; it was a decidedly French classification scheme, nationally popular at the time. The book so impressed the naturalist GEORGE BUFFON that he arranged for its publication by the government and a post for the author as assistant botanist in the Jardin du roi. Although Lamarck continued to add to this work, by 1802 he was involved in other projects and turned the revisions over to an associate to add new species and update the older entries.

The year 1793 was a momentous one in France. The king and queen lost their heads, and the scientific community was involved in the reorganization of all royal institutions that the Revolutionary Tribunal had decreed. Thus, the Jardin du roi became the Jardin des plantes, and within it the collections were renamed. The natural history collection became Le Musée Nationale d'Histoire Naturelle (National Museum of Natural History). While in theory all the resident faculty were professors on an equal footing, in reality the position given to Lamarck in 1800 as professor of zoology was a lowly one. He was to be the expert on insects and worms. These were lumped together as *wormes* by Carolus Linnaeus (the Swedish botanist Carl von Linné), whose classification scheme by then was emerging as the best one available for all organisms.

Lamarck's efforts at sorting out the many parts of the rapidly growing category resulted in the creation of a new field in biology. He created the term "biology" to distinguish living organisms, those that increased in number and complexity, from inanimate things such as rocks. That one could not become the other was a concept that needed clarification, and his idea that living things were composed of cells predated the work of Matthias Schleiden and Theodor Schwann, early 19th-century German microscopists. This botanist and physiologist concluded that all the living cells they examined contained a substance they called protoplasm, which they defined as the "living material" of the cells. Lamarck also created the term "invertebrate" and placed in separate categories Crustacea, Arachnida, Annelida, and Insecta, and also removed barnacles and tunicates from the Mollusca, a group that he exhaustively cataloged. Lamarck spent much time on invertebrate paleontology, beginning by classifying shells with the aim of trying to find a link between modern and fossil forms. If they were similar, he reasoned, then extinction and change must be a factor in development.

Beginning in 1801, Lamarck published a number of works in which he attempted to explain evolutionary change and divergence. Unlike ÉTIENNE GEOFFROY SAINT-HILAIRE, Lamarck did not believe these changes were solely the result of environmental change but a response to it. Thus, Lamarck's First Law was that animals developed structures or organs or increased their size and complexity because of use and environmental conditions. Lamarck's Second Law stated that these changes in animal bodies were inherited. He published his conclusions in *Philosophie zoologique* (Philosophy of zoology) in 1809 and was condemned by his colleague GEORGES CUVIER. The theory of inherited changes was not totally new with Lamarck; another who had similar ideas was Erasmus Darwin, CHARLES DARWIN's grandfather, who had published a work with similar concepts earlier. It seems that the two were unaware of each other's work.

While engaged in invertebrate biology, Lamarck also collected data on weather and geology, and pursued chemistry. In chemistry, he was a defender of the phlogiston theory. Phlogiston was defined as an invisible fluid with negative mass. This substance was proposed as the explanation for why in experiments a metallic sample weighed more than it did before it was burned. For example, on heating, mercury, a silvery liquid,

Jean-Baptiste Lamarck *(E. F. Smith Collection, Rare Book & Manuscript Library, University of Pennsylvania)*

turns into a heavier red powder. This was attributed to a loss of the phlogiston. Lavoisier thoroughly discredited this notion by isolating oxygen and proving that mercury combined with oxygen to produce an oxide (the red powder) that weighed more than the original liquid mercury. Although the idea had been thoroughly discredited by Antoine Lavoisier, the best chemist in France, phlogiston proved to be an attractive idea, and it was accepted for decades. Lavoisier, the discoverer of oxygen and its role in combustion, was executed in 1792 by the revolutionary tribunal during the Reign of Terror. Strangely, Lamarck did not study the physiology of the animals he so carefully cataloged. Although he was never a popular author, he did publish a number of works in addition to *Philosophie zoologique.*

The most significant was *Histoire naturelle des animaux* (Natural history of animals); the first volume appeared in 1815 and the second in 1822.

With his sight failing by 1818, Lamarck spent his last years in poverty being cared for by his daughters; he had been married four times.

Cuvier, a dominant figure in biology and science theory at the time, discredited the evolutionary theories Lamarck had proposed. While respecting Lamarck's work on the classification of invertebrates, Cuvier expended much energy trying to disprove the theory that inherited change was not God's creation. A vindictive man, Cuvier never forgave Geoffroy St.-Hilaire for delivering Lamarck's funeral oration. He saw the eulogy as an acceptance of Lamarck's biology, and any attempt to look favorably on the creator of a scheme that linked the structures and functions of animals to each other and to those of fossils was enough to make the speaker an enemy of Cuvier's. Nonetheless, both Darwin and ERNST HAECKEL, the German physiologist, looked to Lamarck as a forerunner of their own efforts to explain evolutionary theory. Darwin began his own work with attempts to disprove Lamarck but eventually found that it led him in directions that were new, exciting, and troublesome.

When he died, old, poor, and blind, on December 28, 1829, Lamarck was buried in a rented grave and years later disinterred. The body disappeared and was never found.

⊠   **Lang, Arnold**
(1855–1914)
Swiss
Zoologist

Arnold Lang was born in Oftringen, Switzerland, on June 18, 1855, the son of Adolf, a local businessman. He attended local schools and began his academic career at the University of Geneva, where he concentrated on zoology. He soon switched to the University of Jena in Germany and

received a doctorate in zoology there in 1876. Lang's first significant appointment was as Switzerland's representative in the zoological station at Naples, 1878–79. He later became an assistant and stayed on until 1885. This international research station was devoted to marine studies, specifically the flora and fauna of the Mediterranean. After a short period in Jena as *Privatdocent* (a post as research fellow), in 1886 he was, with ERNST HAECKEL's recommendation, appointed professor in the newly created department of phylogenetic zoology at the University of Jena. At that time, Haeckel was the acknowledged biology expert in all of Germany.

Lang returned to Switzerland in 1887. This was a significant year for him; he married Jeanne-Mathilde Bacherlin. Two years later he was appointed full professor of zoology and comparative anatomy at the University of Zurich. Concurrently, he held the same post at the Eidgenössiche Technische Hochschule (Federal Technical University) where he was also director of zoological collections and founder of the Zoological Institute.

Along with his administrative duties, Lang continued the research he had begun while in Naples. Marine worms were his area of interest; he studied Annelida, Turbellaria, and Platyhelminthes, concentrating his efforts on the structure and function of the nervous systems of these simple animals. Recognized for his contributions to science, Lang was a member of the Société des médecins et naturalists de Jassy, the Academy of Natural Sciences of Philadelphia, the Swedish Academy of Sciences, the Royal Society of Sciences of Uppsala, and several other European learned societies.

Following a suggestion by Haeckel, Lang translated JEAN-BAPTISTE LAMARCK's work, *Philosophie zoologique* into German and continued to examine Lamarck's evolutionary concepts. He launched a series of hybridization experiments working with *Helix* (snails), which ultimately confirmed the work of Gregor Mendel, the Austrian botanist, on inheritance of characteristics. As teacher, mentor, careful experimenter, and evolutionary thinker, Arnold Lang made significant contributions to marine science. His popular text, *Lehrbuch der vergleichenden anatomie der wirbellosen tiere* (Textbook on comparative anatomy of invertebrate animals) was published in sections between 1888 and 1894. The material was later translated into French and English. Profoundly depressed by the start of the Great War, he died in Zurich on November 30, 1914.

Lankester, Sir Edwin Ray
(1847–1929)
British
Zoologist

Ray Lankester was one of the great "characters" in British science. This forceful personality was born in London on May 15, 1847, the son of Edwin Lankester, a popular doctor who was also serious about scientific study. The younger Lankester was educated at the prestigious St. Paul's School. While there, he showed his scientific talent early; he published a paper on Pterapsis, a fossil fish or reptile, when he was 16. He then went on to Downing College, Cambridge, but left for Christ Church College, Oxford, to finish the work in geology and zoology for his university degree in 1870. He then proceeded to the universities in Vienna and Leipzig to attend the lectures of famous scientists. After a period as Britain's representative at the research station in Naples, Lankester returned to Britain when he was appointed fellow of Exeter College, Oxford, in 1872.

As mentor and teacher, he directed the careers of a great number of students. His research interests ranged through just about every phylum extant and extinct. He systematized the study of embryology, inventing several indispensable words, such as blastophore and invagination. Stirred by his experiences in Naples, Lankester

was the founding spirit and mover for the Marine Biological Association's research laboratory in Plymouth. This facility opened in 1884 as a training base for British naturalists. He left Oxford when he was appointed professor of zoology at University College, London, and returned to Oxford in 1891 as Linacre Professor of Comparative Anatomy.

As a member of the circle of CHARLES DARWIN's devoted friends and ardent supporters, Lankester gave public lectures supporting Darwin's work. He was also an outstanding critic of JEAN-BAPTISTE LAMARCK's theories of inheritance of acquired characteristics. All this occurred before the theories of the botanist Gregor Mendel, who saw genes as transmitters of inheritance, were widely known. He was elected a member by the Paris Academy of Sciences in 1899 and awarded its annual medal for excellence in research by the Royal Society.

Ray Lankester again left Oxford for London in 1898, when he was appointed director of the Natural History Museum. By then, that institution had been separated from the British Museum. Lankester hated the administrative detail involved in being director and retired with a knighthood in 1907. In retirement, he wrote several books on scientific subjects directed at lay readers: *Treatise on Zoology* (1900–09), *Science from an Easy Chair* (1910–12), which was based on his newspaper column, *Extinct Animals* (1909), *Diversions of a Naturalist* (1915), and *Great and Small Things* (1923).

Lankester was a marvelous teacher, researcher, director, and enthusiast. He wanted everyone, from the serious student examining and classifying a specimen to a schoolchild in the museum, to love the subject and be fascinated by it. Another contribution he made to the scientific literature was his service from 1869 to 1920 as editor of the *Quarterly Journal of Microscopial Science*, a journal that had been established by his father in 1860. Ray Lankester died in London on August 15, 1929.

Lenz, Heinrich Friedrich Emil
(1804–1865)
Russian
Physicist

Heinrich Friedrich Emil Lenz, also called Emil Khristinovich, was born on February 12, 1804, in Dorpat, Russia, now Tartu in Estonia. His father was a local government official who died in 1817. The boy was raised by his uncle and mentor, E. F. Giese, who was the chemistry professor at the University of Dorpat. Lenz attended the university, where his teachers were his uncle and G. F. Parrot, the founder of the physics department. Parrot was later rector, the chief academic officer of the university. Both professors urged Lenz to participate in August Friedrich Kotzebue's second scientific voyage around the world. Kotzebue was an admiral who had been commanded by the Russian government to sail around the Pacific and see what Russia could take that the English, French, Dutch, Spanish, and Portuguese had not already laid claim to. Lenz, at 19, left on the *Predpriatie* on a voyage of over two years.

His specific duties while on board involved using instruments designed for him by Parrot. He measured the temperature and specific gravity of sea water. Specific gravity is the ratio of the mass of a body to the mass of an equal volume of water at the same temperature. He was comparing seawater samples to pure water. Samples were taken from varying depths from the surface down to 2 km (3,280 ft), and his accuracy in measurement equaled that of the later voyage of the HMS *Challenger*, the British study expedition (1872–76). Studies of ocean water samples showed that the salt content of ocean water decreased as a ship approached the equator in either hemisphere, and also that the salinity of oceans varied. The Atlantic salt load was different from that of the Pacific, and both differed from the Indian Ocean. The Atlantic was the saltiest. Lenz also took note of the temperature differences between air and surface water at varying latitudes.

After his return in 1825, Lenz presented his results to the St. Petersburg Academy of Sciences,

and in 1828 became the scientific assistant in the academy. His career was based in the capital from then on. A month-long trip to southern Russia and the Caucasus, where he examined geomagnetism in 1829–30, was followed by his appointment as associate Academician in 1830 and then full Academician in 1834. He is well known for his research on electrical conduction and electromagnetism. He discovered the principle known as Lenz's Law, which states that an electric current induced by a changing magnetic field will flow in a direction opposing the effect that created the change. These opposing fields occupying the same space at the same time result in a pair of forces. These forces are felt when you turn on a generator and generate electricity. This is a fundamental concept in the conservation of energy.

Lenz continued his research on the nature of electrical resistance to the end of his life. He suffered a stroke and died in Rome while on vacation on February 10, 1865.

Lesueur, Charles-Alexandre
(1778–1846)
French
Natural Historian

Charles-Alexander Lesueur was born in Le Havre, France, on January 1, 1778. His father, an officer in the French Admiralty, enrolled him in the *École Royale Militaire* (Royal Military School) in Beaumont-en-Auge when the boy was nine. After graduating, he served for two years (1797–99) as an officer in Le Garde Nationale at Le Havre. It was there that he had a chance to join the French exploratory voyage to Australia and Tasmania. The expedition was mounted because the English were sending Matthew Flinders on a similar voyage. The French ships were two corvettes, *Géographie* and *Naturaliste*, commanded by Nicolas Baudin. The departure in 1800 was the start of an almost four-year-long trip in which Lesueur started as assistant helmsman and became the artist-naturalist for the official naturalist aboard, François Peron.

Lesueur's magnificent drawings were, for the most part, unused in the final official report.

Unemployed for a while after his return to France in 1804, and teaching drawing thereafter, Lesueur had a chance meeting with William Mc-Clure in 1815. Lesueur packed his drawings and accompanied the rich American to the West Indies, where they explored the new territory's fish, corals, and other exotic sea life. By 1816, Lesueur was in New York, and setting out from that point, he traveled around the northeast collecting plants, animals, and fossils. Settling in Philadelphia, Lesueur spent nine years there as a naturalist, engraver, and drawing teacher. He was also a member of the American Philosophical Society and the Academy of Natural Sciences of Philadelphia.

Traveling again, he joined Robert Owen's utopian community in Indiana in 1825, and stayed on after that venture failed. He had begun to prepare a large work describing North American fish. The work was never finished but did entail a great deal of traveling. Again unemployed, Lesueur returned to France in 1837 and spent more than eight years in Paris applying repeatedly for a suitable position. Finally, he was appointed curator, in 1846, of the newly opened Muséum d'Histoire Naturelle in Le Havre but it was a short-lived career. He died in his native city on December 12, 1846.

He was the first reporter on the paleontology of the Mississippi River valley and left a legacy of his magnificent drawings. The drawings are still in Le Havre having survived the destructive bombing of both world wars.

⊠ Leuckart, Karl Georg Friedrich Rudolf
(1822–1898)
German
Parasitologist, Zoologist

Karl Leuchart was born in Helmstedt, Germany, on October 7, 1822. He attended the gymnasium

(secondary school) there and was drawn to the work of his uncle Friedrich Leuckart, a professor of zoology at the University of Freiburg. As there was no department of natural science, young Karl enrolled in the faculty of medicine at the University of Göttingen, where he met the zoologist, Rudolf Wagner. When Leuckart finished his medical degree and the licensing examinations necessary for practicing medicine in 1845, Wagner took him on as his assistant. Rising rapidly to zoology lecturer by 1847, Leukart embarked on a collecting trip to study the invertebrates of the North Sea coast. In the next year, the chemist Justus von Liebig and the anatomist Theodor Bischoff were attracting a young faculty to the new University of Giessen. They invited Leuckart to join them in this new institution; he did, and stayed there until the University of Leipzig offered a better position in 1869. Well pleased with him and his research, the University of Leipzig built a new zoological institute for Leuckart in 1880.

Honored by the scientific community in Giessen, Leuckart was awarded an honorary Ph.D. by his home institution in 1861. The Ph.D. was a new honorific at this time. It was given by a university in recognition of continuing scholarship almost as honorary degrees are given now.

After his move to Leipzig he was a privy councilor to the king of Saxony, and a member of several international scientific societies, as well as a medallist. He also served as rector (chief academic officer) of the University of Leipzig for the 1877–78 academic year.

Well known in Europe, Leuckart made his reputation in the 1840s. In 1847 and 1848, he began working on the separation of the Radiata (animals with five-part symmetry), a group identified earlier by the influential French naturalist, GEORGES CUVIER, into two groups that he named Coelenterata and Echinodermata. Leuckart divided the so-called Metazoa into six phyla, Coelenterata, Echinodermata, Annelida, Arthropoda, and Vertebrata. This work earned him much praise, but it was also the subject of considerable controversy. Old ideas, particularly those of Cuvier, had great weight and were not easily set aside.

Using physiology as a means of classification, Leuckart began an extensive study of parasitic worms in 1852. He was best known to his contemporaries as the man who produced the outstanding work *Trichina* (1853), on the parasitic worms of the same name. These parasites affect cattle and particularly swine, making consumption of the infected meat a danger to human health. Isolation and identification of these organisms made a huge difference in public health. Although he never finished his definitive two-volume work on human parasites, it nevertheless became a classic in the field. He was still working on his masterpiece when he died in Leipzig on February 6, 1898.

⊠   Loeb, Jacques
(1859–1924)
German/American
Physiologist

Jacques Loeb was born in Mayen, Prussia, now Germany, on April 7, 1859. His parents were Benedict, a wealthy importer, a philosopher, and a Francophile, and Barbara Isay. Jacques Loeb's younger brother Leo became a well-known doctor at Washington University in St. Louis, Missouri. Jacques's original intent was to study philosophy, his interest being the determination of freedom of will. He began his studies at the University of Berlin in 1880 but was quickly disappointed by the didactic approach of the philosophy department and determined to study "will" scientifically. He then enrolled at the University of Strasbourg to study localization of brain function and was granted a medical degree in 1884. He considered his education a five-year-long waste of time. In 1886, he moved to Würzburg to study with Adolf Fick who was then trying to apply physics to biology. Loeb served as

Fick's assistant. His friend Julius Sachs was working on plant tropisms (response to stimuli) and that inspired Loeb to look for similar responses in animals.

A stay at the Marine Biological Station in Naples in 1889–90 brought Loeb into contact with a group in which the dominant intellectual pursuit was that of Wilhelm Roux, the mechanics of evolution. Everyone who aspired to a research career in biology dreamed of a period of study at the Naples Biological Station. Since many scientists from many countries went there, it acted as a continuous seminar session for exchange of ideas between people who otherwise had far less chance of working together. One of the people in the group was Thomas Hunt Morgan, later the famous geneticist at Columbia University. He and Loeb were friendly rivals in future years.

In 1890, Loeb married Anne Leonard, an American philologist who had been working in Zurich for her doctorate. They moved to the United States the following year; Loeb, a dedicated liberal, believed Germany under Bismarck was an autocracy he had no place in. His first professional appointment, as a teaching and research professor was at the new women's college in Bryn Mawr, Pennsylvania. Short stays at the University of Chicago and University of California in Berkeley were followed by an appointment in 1910 at the Rockefeller Institute (now Rockefeller University), the research facility in New York City, where he did his most significant work.

Beginning in 1888 Loeb worked on what he called animal tropisms. He worked first on caterpillars and their response to light. A prominent evolutionary zoologist, Herbert Spencer Jennings, disagreed and stated that this was a response to stimulus; he called it instinct. Loeb reexamined his thinking and decided that he had exchanged one mystical concept for another, discarded both and set himself to find a measurable cause of the response. Animal tropism, Loeb's first idea of response to light, was a mechanistic answer to why animals respond to stimuli. Loeb then thought that the replacement of one word (tropism) that was unexplained by another word (instinct), also unexplained, added nothing to understanding why there was a response. There had to be a physiological reason.

He began to experiment on the effect of salts on developing eggs. Exploring the very beginnings of life, Loeb exposed fertilized sea urchin eggs in seawater to additional salt. Thomas Morgan did the same to unfertilized eggs. Loeb achieved segmented eggs that developed into larvae; Morgan's eggs segmented and did not produce larvae. The effect of the osmotic pressure on the embryos made a difference. Loeb experimented using a variety of salts with varying effects, showing that environment made a difference in egg maturation. Osmosis is the flow of water through a semipermeable membrane such as that of a cell. If the concentration of salt outside the membrane—in the environment—is greater than that within the cell, water will flow from the cell into the surrounding watery environment to equalize the salt concentration. If the salt concentration in the environment is less than that within the cell, the cell will allow water to flow in. In the first case the cell may become so dehydrated it cannot function and in the latter it may burst.

Beginning in 1899, Loeb explored the possibilities of parthenogenesis (asexual reproduction) and continued to study that aspect of embryology, publishing his results in the book "Forced Movements, Tropisms and Animal Conduct" in 1918.

Leaving the study of whole organisms, Loeb studied their component compounds. From 1918 until he died, Loeb worked on protein research and published part of his research in "Proteins and the Theory of Colloidal Behavior," published in 1922. He was well known to the general public, as well as the scientific community, and the model for the senior scientist in Sinclair Lewis's novel *Arrowsmith*. Jacques Loeb died in New York on February 11, 1924.

⊠ Lovén, Sven Ludwig
(1809–1895)
Swedish
Biologist

The son of Christian Lovén, the mayor of Stockholm, Sweden, Sven Lovén was born in that city on January 6, 1809. He attended the University of Lund in southern Sweden, where his mentor was Sven Nilsson, a well-known zoologist. In 1826 they traveled to Norway (then part of Sweden) to observe birds, and Lovén's observations on this journey became his first publication. For another degree, the young man left Sweden for Germany and studied with CHRISTIAN EHRENBERG, the leading zoologist in Germany, who persuaded him to concentrate on marine biology. The first significant publication in marine science that Lovén published, in 1835, dealt with microscopic crustaceans. He continued to study plankton for the rest of his scientific career, tracing the life cycles of these tiny, newly discovered marine organisms.

Another collecting trip, this one 17 months long (1839–40), took him to Arctic waters north of Norway to collect plankton samples and study Paleozoic fossil forms. After he was appointed curator of invertebrates at the Museum of Natural History in Stockholm in 1841, he began a major study of mollusks, the shelled invertebrates that include clams, oysters, scallops, limpets, and snails. Radulae, the rasping mouthparts of mollusks, are truly distinctive features in Gastropoda (snails), a very large class. Lovén used common structures like that as indicators for classification, leading to the publication of *Index Molluscorum* in 1846; this was the definitive work at that time on all Mollusca. Another work that followed shortly was a detailed study of the radulae.

In collaboration with Friedrich Muller, the physiologist, Lovén discovered polar bodies, structures in cell maturation and division that are common to all animal cells. While not restricted to marine science, this was a major effort in understanding how a cell grows, matures, and reproduces. While working on this, Lovén continued to study the metamorphosis of bivalves.

The study of fossil bivalves led Lovén to postulate that most of the Scandinavian peninsula was once under water. He then used fossils to track climate change, ice movements, and landform change and elevation. He concluded that the land is still springing up from the crushing weight of the last glaciation.

Lovén's classic work on Echinoderm anatomy was published in 1874. He continued to work, mentor students, and coedit the journal that he had founded in 1844 with Jöns Jacob Berzelius, the premier chemist in Sweden, *Summary of the transactions of the royal academy of science*. Berzelius was known throughout Europe for his discovery of several elements, and for a fundamental textbook in chemistry. Lovén retired from teaching in 1892 and died in his Stockholm home on September 3, 1895.

⊠ **Lyell, Sir Charles**
(1797–1875)
British
Geologist

Born in Kinnordy, Scotland, on November 14, 1797, Charles Lyell, was the oldest of the 10 children of Charles Lyell, a noted botanist, and Frances Smith. Neither parent was from Scotland: both were originally from the London area and both families had strong connections with the Royal Navy. The family moved to Hampshire when the future geologist was very young; there his father collected plant specimens, and the child amassed a beetle collection.

The young Lyell began his education at Oxford University when he was about 17. His first interest was mathematics, but the lectures in geology given by William Buckland, a famous uniformitarian, so interested him in the subject that

he made it his special interest. The prevailing geological theories of the 1820s were those of the uniformitarians, who included James Hutton and Roderick Murchison. Unlike their predecessors, the neptunists, who believed that the Earth had in ancient times been covered by a vast ocean, the uniformitarians thought that Earth changed only gradually. In their view, all geologic processes were slow and took eons. Catastrophists, on the other hand, explained the disappearance of old species and the introduction of new ones as a continuous round of total devastation followed by new creation. He also continued his studies in preparation for the bar.

Troubled by eye problems, Lyell left Oxford to convalesce, and his father took him traveling in Europe in 1819. On this first of many trips abroad, Lyell did some geological exploring. In the same year he joined the Geological Society of London and the Linnaean Society. Later he entered Lincoln's Inn, one of the associations of barristers serving in the law courts of London, to prepare for entry into the legal profession. Studying slowly to give his eyes as much rest as needed, he spent much of his time visiting friends in the southeast of England. This gave him the opportunity to observe the similarities of the rock formations in the southeast and on the Isle of Wight.

In 1823, on another trip, Lyell met the great men of science in Paris among them WILHELM VON HUMBOLDT, GEORGES CUVIER, and most particularly, Pierre Prévost, a noted Swiss physicist, who had been JEAN-BAPTISTE LAMARCK's student and then collaborator. He was a geologist and, unlike Cuvier, not a catastrophist in his explanations of the history of the Earth. Lyell studied the geology of the Paris region with Prévost.

The 1820s were a busy decade for Charles Lyell. In 1823, he was secretary of the Geological Society. Finally admitted to the bar in 1825, Lyell practiced law for about two years. The court moved from one place to another hearing cases, and this travel throughout southern England afforded Lyell the chance to do more "geologizing" (sic). He enjoyed his time as a traveling barrister; it gave him the opportunity to visit different regions and explore the rock formations wherever he went. Recognized as a serious geologist by the Royal Society, he was elected a fellow in 1826.

His exploration while traveling continued in 1828, when he accompanied Roderick Murchison and his wife to France, where Lyell found young geological landforms that were analogous to older formations he had identified in England. The party went on to Italy where they examined the temple in Pozzuoli, which originally had been elevated but had later subsided in historical times. Lyell concluded that sudden Earth movements were indeed possible and wrought dramatic changes in Earth's conformation. Finds in Sicily led him to believe that rock formation was not just a function of time but of the prevailing conditions, and that these could repeat themselves. Extinctions did occur, and they were the result of numerous factors, not just a single cataclysmic event. Lyell's work combined the idea of uniform processes such as erosion shaping the Earth, and sudden events, such as earthquakes creating rapid change; both occur, both have effects that are obvious in the fossil record.

Lyell produced the first volume of his classic work, *Principles of Geology,* in 1831. It was an instant success and one of the most precious objects taken aboard the *Beagle* by CHARLES DARWIN. The subsequent two volumes were published by 1833. The book was a continuing work because so much new evidence was emerging from a large number of researchers in many places. Continuous updating meant that eventually 12 editions appeared, the last one after Lyell's death.

Lyell taught geology at Kings College, London, from 1831 to 1833, and married Mary Elizabeth Horner in 1832. She was the daughter of Leonard Horner, a fellow geologist. Mrs. Lyell became a noted conchologist, a specialist in molluscan shells, and she was Lyell's companion on his travels. Although Lyell had given up teaching, he continued to travel and work on his famous book and others, such as *Elements of Geology,* published in 1838. This work was a description of European rock forms and fossils from earliest geologic time to the present. The separation of Cretaceous from Tertiary, and the division of the latter into Eocene, Miocene, Pliocene, and Pleistocene Ages was Lyell's work, and the terms used to describe these ages were his.

Established as a great authority in geology, Lyell traveled and lectured extensively. He visited Europe quite often and North America four times in 1841–42, 1845, 1852, and 1853. The diaries of the first two voyages were published, the first as *Travels in North America* in 1845, and in 1849, *A Second Visit to the United States.* Both were scientific and popular successes. The other two voyages, in 1852 and 1853, were relatively short visits. Honored by his nation, Lyell had been knighted in 1848 and created a baronet in 1864, the year that he was president of the British Association for the Advancement of Science. Recognizing Darwin as a formidable researcher, Lyell was one of a group of scientists who in 1856 visited Darwin in Downe and urged him to finally publish the works that were the result of his round-the-globe voyage. Although Darwin had relied heavily on Lyell's work, at first Lyell did not agree with Darwin about evolution. In the work *The Antiquity of Man,* first published in 1863 and revised four times before 1873, Lyell continued to assert that evolutionary theory could not apply to man. However, Lyell declared himself convinced by most of Darwin's work in 1864, on the occasion of Darwin's receiving the Copley Medal from the Royal Society.

At the end of Lyell's life, a controversy arose with William Thomson, Lord Kelvin of Largs, an outstanding theoretical physicist. Among his works were the establishment of the temperature scale that bears his name, the definition of absolute zero temperature, the development of thermodynamics, and the interrelationship of forms of energy. The question was the age of Earth. Lyell had proposed a figure of about 4 billion years and Kelvin argued with him, basing his ideas on the length of time required to cool a body of Earth's size and density. Kelvin thought that Lyell's estimate was far too long. However, Kelvin did not know of radioactivity, the heat source that explained how the Earth stayed hot, which made Lyell's estimate, based on geologic processes, the one that was correct after all.

In spite of the eye problems that eventually caused his blindness, Lyell continued to work, correspond with other geologists, and speak on his continuing interests. The death of his wife in 1874 was a great blow, and he died months later in London on February 22, 1875. He is buried in Westminster Abbey.

M

Maillet, Benoît de
(1656–1738)
French
Geologist, Oceanographer

Benoît de Maillet was born on April 12, 1656, into a family of the lesser nobility in Lorraine. He served the French monarchy as consul in Cairo from 1692 to 1708. His later appointments took him to Leghorn (Italy), to the Levant, and to the Barbary States in North Africa. When he retired to Marseilles, he wrote of his extensive Mediterranean travels in *Descriptions de l'Egypte* (Descriptions of Egypt), which appeared in 1735 and was widely read.

The prevailing theory of the history of the Earth in de Maillet's time was neptunism; in his memoir he used that theory. His explanation of the geological history of Earth proposed a universal ocean that eventually shrank to reveal the landforms now visible. Benoît de Maillet was an early advocate of the theory that Earth's age dated back for billions of years. As his study continued, he took on the beliefs of a uniformitarian geologist, in which one assumes that the geological layers of the earth were not formed by divine intervention, but by natural phenomena (such as erosion, flooding, and earthquakes). He went on to propose a history of transformation of sea creatures. This was not exactly evolution; the transformations were the result of changing environments. Maillet's influence on GEORGE BUFFON and GEORGES CUVIER was considerable. He was a starting point for both in their thinking about the origins of the Earth and the creatures that were on it. Maillet's theory of gradual change that was directed by environment was a direct precursor of JEAN-BAPTISTE LAMARCK's theories of evolution.

Makarov, Stepan Osipovich
(1849–1904)
Russian
Oceanographer

Born in Nikolayev on March 8, 1849, Stepan Makarov was the youngest of the five children of Osip Fydorovitch, a retired naval officer. The boy was placed in the Royal Naval Cadets' School when he was 10 and he graduated in 1865. After serving as an ensign, he commanded a steamer in the Black Sea in the war with Turkey (1881–82). After the war, he sailed on the *Tamen* while attached to the Russian embassy at the Turkish court. He was occupied in data collection in the Bosporus where his field of study was current mapping. There is a famous current phenomenon in the Bosporus; a surface current is obvious but there is also a deep and much faster countercurrent present. This was known earlier and written of by LUIGI MARSIGLI, a 17th-century natural scientist. Makarov invented a device to measure the differing velocities of the two currents, which he also used while collecting information about the variation in density of water in the Black Sea, the Bosporus, and the Sea of Marmara. This work was published as *On the Exchange of Water of the Black and Mediterranean Seas* in 1885.

After much urging, the Russian government's official voyage to explore the deep ocean was approved; Makarov was given command. A sense of national pride drove this expedition. All other major nations were engaging in marine research; therefore the Russian scientific community believed that a Russian presence in research at sea was required. The ship *Vitiaz* went around the world, measuring water densities and temperatures between depths of 25 and 800 meters (80 and 2,750 feet). Their measurements constituted the first temperature tables of the northern Pacific Ocean. The ship also added to every nation's knowledge of the origin of deep Pacific Ocean water, the patterns of ocean currents, and the effect of the Coriolis force on currents. The Coriolis effect is a part of the phenomenon of right-hand drift in the Northern Hemisphere. The rotation of the Earth from west to east pushes all fluids (water and air) before it, affecting both ocean currents and weather. The publication of this work in Russian and French brought Makarov international attention.

Later, on the *Ermak*, an icebreaker, Makarov attempted to open a permanent passage north of Siberia in order to study the temperatures and salinity of water from various Arctic locations. This project was ultimately not successful, but he spent three summers in the period 1899–1901 trying to achieve this goal. His naval duties made further research impossible. Makarov died on April 13, 1904, when his battleship was blown up in Port Arthur during the Russo-Japanese War.

⊠   **Mann, Kenneth H.**
(1923–   )
British/Canadian
Marine Ecologist

Born in Dover, England, on August 15, 1923, Kenneth Mann was educated locally and received the B.Sc. degree in biology from University College of London in 1949. He continued his education in zoology and received a Ph.D. from the University of Reading in 1953 and a D.Sc. in 1966. His career developed in Reading, where he was lecturer in zoology (the American equivalent position is assistant professor) and was also attached to the Marine Ecological Laboratory of the Bedford Institute of Oceanography. A further appointment

Kenneth Mann *(Courtesy Dr. K. H. Mann)*

brought him to a chair in Dalhousie University in Halifax (Nova Scotia), where he is professor emeritus.

Mann's research interests are the biophysical interactions in the sea, aquatic ecosystems, and food chains. In the 1980s and 1990s, his research was centered on the ecosystem of the lobster population of the North Atlantic. In spite of the long history of lobster fishing in this area, until these studies began, scientists knew very little about the life cycle of these bottom-dwelling arthropods. The second edition of Mann's book, coauthored by J. R. N. Lazier and published in 1996, is a summation of some of his research. Mann was editor of *Journal of Animal Ecology* and is a member of the American Society for Limnology and Oceanography and the British Ecological Society, and a fellow of the Royal Society of Canada. He is a Research Scientist Emeritus at the Bedford Institute of Oceanography. He is now engaged mostly in writing on coastal ecosystems, especially those dominated by seawater, seagrass, and mangrove production. A particular interest is in the relationship between physical oceanographic forces and biological production.

⊠ Marion, Antoine-Fortuné
(1846–1900)
French
Botanist, Zoologist, Geologist

Antoine-Fortuné Marion, born on October 10, 1846, was a Provençal and a classmate of Émile Zola, the author-journalist, in the *lycée* in Aîx. In 1862, the professor of geology at the Faculté des Sciences in Marseilles chose Marion as his assistant in natural history before he had finished the work required for a baccalaureate in arts and letters. Marion finished that degree in 1866 and the degree in science in 1868. The work he presented to the Faculté des Sciences for his doctoral degree was a study of the anatomy and zoology of non-parasitic segmented marine worms.

An appointment as director of natural science in the Marseilles *lycée* gave Marion a laboratory and a public presence. From that position, he promoted a marine laboratory, which was finally opened in 1878 with financial support from the city of Marseilles. Two years later, Marion became director of the Natural History Museum in Marseilles and founded its publication, *Annales*. In his lifetime, this journal was almost exclusively a marine science publication.

While still a child, Marion had given the Marquis de Saporta a present of a fossilized magnolia leaf. They remained friends and coauthors of botanical works as long as Marion lived. He was also involved in the national struggle against *Phylloxera*, the aphid-like insect that was devastating the vineyards of France. The work he began with his doctoral studies continued. He went on to examine free-living roundworms of the Mediterranean, and then to study very small, albeit multicellular, free-living marine organisms, the rotifers in 1872, nemerteans from 1869 to 1875, zoantharians in 1882, and alcyonarians. The work on the latter appeared in several publications in 1877, 1882, and 1884. He returned to study segmented worms again in 1874, and then worked on crustacean parasites in 1882, when he turned his attention to the phyla Enteropneusta and Mollusca in 1885–86. Enteropneusta is a phylum of worm-like, small marine creatures. In Marion's time they were considered worms. Mollusks are clams, oysters, scallops, snails, animals much larger and more complex than the Enteropneusta, whose classification is still uncertain.

Honors were awarded to him for his work on *Phylloxera* by both French and foreign governments. He was also given two prizes by the French Academy of Sciences in 1885 for works involving the dredging of the Mediterranean coast. This dedicated researcher was an early advocate of marine environmental studies and engrossed with his work and its implications. He died suddenly in Marseilles on January 22, 1900.

⊠   Marsigli, Count Luigi Ferdinando
(1658–1730)
Italian
Natural Historian

Count Luigi Marsigli was born in Bologna on July 20, 1658, into a family of the lesser nobility of northeastern Italy. The family name of this student of natural history is also spelled Marsili, and little of his early life is known prior to his appearance in the Austrian army. At the time, much of northern Italy was part of the Austrian empire. While the young officer never had university training, his army career, which extended to 1704, took him to many places on the empire's business.

One memorable voyage was his passage from Istanbul (Europeans continued to call Turkey's capitol Constantinople, its name before the Turkish conquest in the 15th century) to Amsterdam by ship. He was a careful observer and collector of mineral specimens, as well as an experimenter, and this trip gave him perspective on the nature of the Mediterranean and the ability to compare it with the Atlantic and then the North Sea.

Marsigli was a geologist who worked on the question of how the sea became salty and where water in the mountains came from. He created a reasonable explanation of the water cycle, the construction of mountains, and their erosion due to wind and water. In 1681, he published a careful work on the Bosporus, which significantly predated the work of STEPAN MAKAROV, the Russian oceanographer who made a detailed study of the Black Sea in the 1880s. Marsigli was the first to report the lower countercurrent in the Bosporus and remarked on the differing densities of the layers of water in it.

The work that made Marsigli famous, *Histoire physique de la mer* (Physical history of the sea), was known before it was finished in 1724, since he had written to many other European scholars before publication. The encyclopedic work discusses the sea and its relationship to the surrounding territories, the variation in water temperatures, the densities and the chemical contents of various bodies of water. It also discusses the distinct currents and wave action.

The biology of sea life is a considerable part of his work. He had an early appreciation of the wealth of organisms in marine waters; earlier writers were so fixed on the plants and animals on land that they barely noticed those in the waters. One of his errors was that he classified coral as plants. In his defense, however, this classification was an improvement over earlier natural historians who called them nonliving rocks.

Marsigli's most detailed work on a single subject—a careful study of Lake Garda, an Italian alpine lake—was not published. The rough outline and notes survived. In his research, he considered its water sources, mineralogy, geology, flora, and fauna profile. Marsigli was elected a member of the Royal Society of London in 1722. His work on the Mediterranean was a major source for further research directed by the British Admiralty. The British were much more interested in the Mediterranean, its tides and currents, after the foundation of the Royal Society of London by Charles II in 1662. The society then sponsored work to chart all ocean currents and they used whatever sources they found. Marsigli was a primary source. After the English took Gibraltar in 1704, the interest in the currents and water flow through the strait occupied several of the members of the society who used Marsigli's methodology.

In 1712 Marsigli had organized the Academy of Science of the Institute of Bologna and continued to be an active member for the rest of his life. He died in Bologna on November 1, 1730.

⊠   Matsuyama, Motonori (Matuyama)
(1884–1958)
Japanese
Geologist, Geophysicist

Born in the town of Usa, in the Kyoto Prefecture on October 25, 1884, Matsuyama, the son of a Zen abbot, adopted the name of his in-laws when he married in 1910. His academic career began in Hiroshima Normal College with a degree in 1907. A normal school was a two-year program to train primary or secondary school teachers. The teacher training school at Hiroshima has since expanded and is now the University of Hiroshima. After teaching physics and mathematics for a year, he was accepted at the Imperial University in Kyoto and graduated in 1911 with a doctorate in physics. Geophysics was a logical continuation of that work, and his mentor was Toshi Shidu, with whom he investigated gravitation. The 1912 work that Matsuyama coauthored with Shidu was *On the Elasticity of the Earth and the Earth's Crust*.

By 1913, Matsuyama was appointed lecturer at the Imperial University, and in 1916, he became assistant professor at the Geophysical Institute. His 1918 dissertation, published as *Determination of . . . Gravitation Potential on Jaluit Atoll*, investigated the thickness of specific layers and geological profiles of the coral atoll, useful in the exploration for mineral resources. By 1919, Matsuyama was doing experimental geology in the United States at the University of Chicago. There, he made laboratory models of glaciers and by subjecting them to pressure and other variables, studied the deformations in ice.

On his return to Japan in 1921, he was appointed professor of theoretical geology at the Imperial University. His research from 1927 through 1932 was a continuation of his earlier work on gravity determinations extended to other regions, notably Korea and Manchuria. Using instruments perfected by FELIX VENING MEINESZ in 1934 and 1935 in a submarine, Matsuyama explored the Japan Trench, a deep near the Japanese home islands, and other oceanic abyssal regions near the Caroline and the Mariana Island groups. He found free-air gravity anomalies in both of these tropical Pacific island areas. A gravity anomaly is the difference between measured gravity at sea level and predicted gravity on a reference ellipsoid. The next year he led a gravity survey of the Japan Trench; this significant oceanic canyon is the site of frequent seismic activity that produces earthquakes. These undersea quakes trigger tsunamis. Later, a student of Matsuyama's correlated the earthquakes with gravity anomalies.

The Earth is not round. It is flattened at the poles and bulges at the equator, forming an ellipsoid. A reference ellipsoid is one where the only factor used to predict the gravity at a particular point is geographic position. If there is a difference between the theoretical reference point and the measured one, that is a gravity anomaly, an indication that another factor—an unknown—is influencing the gravity.

Matsuyama first started studying coral reefs in 1915 when he remarked on the uplift in the Marianas and the subsidence in the Marshalls. This was followed by a recommendation to the Japanese Association for the Advancement of Science that seabed shift and its angle are areas that should be studied. A research station in the islands was established to do this, but the research conducted there was never published.

Later, this prolific researcher investigated the magnetic field of Earth. In this research, which used some of the findings of Karl Gauss, a 19th century mathematician whose work was applied to a number of fields, including the lines of force in terrestrial magnetism. This topic had been studied by earlier researchers, including EDMOND HALLEY. In this work Matsuyama proposed that the Earth's magnetic field was essentially reversed in the Carboniferous and Tertiary periods and very possibly in the Miocene and Quaternary Eras as well. Matsuyama's most significant work in this area is *On Direction of Magnetization of Basalt in Japan, Korea and Manchuria*. This proposal was later used as a starting point by DRUMMOND MATTHEWS and FREDERICK VINE. Matsuyama retired in 1944 and became professor emeritus in 1946. However, that was not the end of his career. He became president of Yamaguchi University in 1949 and a fellow of the Japanese Academy. He continued at the University until he died of leukemia in Yamaguchi on January 27, 1958.

⊠   **Matthews, Drummond Hoyle**
(1931–1997)
British
Geophysicist

Drummond Matthews was born in London on February 5, 1931, and attended the Bryanston School in Dorset. Entering Cambridge's Kings College in 1950, he received a B.A. in 1954, an M.A. in 1959, and a Ph.D. in 1962; all his degrees were in geology, and he was connected to Cambridge University for his entire career. His first research appointment was to the Geological Falkland Island Survey, from 1955 to 1957. While still working on his doctorate, he was named a research fellow in Kings College and to the department of geophysics in 1960. As his research continued, Matthews rose to assistant director of research in that department in 1966 and to reader—the equivalent of associate professor—in the marine geology department in 1971.

By then, Matthews and one of his doctoral students, FREDERICK VINE, had become geophysical celebrities. Their joint paper in *Nature* (1963) showed the reversals of the magnetic poles in the abyssal rock of the ocean bottom. This work was a continuation of the HARRY HESS theory of continental drift, which postulates that the seafloor consists of new rock emerging from the ridges in mid-ocean, and that of MOTONORI MATSUYAMA, who showed that the magnetic field of the Earth has reversed itself several times in the past. The North Magnetic Pole became the South Magnetic Pole and back again. Matthews and Vine proved that the magnetic particles in abyssal rock are aligned with the magnetic field of the Earth, as it was when those rocks were molten. The effect is a zebra-striped pattern. Rocks from the sea bottom are difficult to bring to the surface. When this was done and the rocks became available to the researchers, Matthews and Vine showed that not only was Hess correct in his hypothesis that seafloor was produced at plate edges as molten rock emerged from deep

within the Earth, but that the iron crystals in this molten, iron-rich rock aligned themselves with the North Magnetic Pole. Since Matsuyama postulated change in polarity, the crystals would have to be aligned differently to reflect this direction change. Matthews and Vine showed that this was indeed the case.

For this and other continuing work on the geomagnetism of Earth, Matthews was elected a fellow of the Royal Society in 1974. He was scientific director of the British Institution Reflection Profiling Syndicate at Cambridge University from 1982 to 1990 and a Fellow of Wolfson College (Cambridge) from 1980 to 1990. He was the recipient of the Balzan Prize in 1982. After retiring to a small town in Somerset, Matthews died of a heart attack on July 20, 1997, in Taunton.

⊠   **Maury, Matthew Fontaine**
(1806–1873)
American
Oceanographer, Physical Geographer

Matthew Maury, son of Richard and Diana (Minor) Maury, who were both of old Virginia families, was born on January 14, 1806. He was their seventh child and educated locally. In 1825, he joined the U.S. Navy as a midshipman, following the example of an older brother, and made three long voyages between 1825 and 1834, one to England and Europe, another around the world, and the last to the western coast of South America. His first publications are based on his days on active sea duty; *On the Navigation of Cape Horn* (1834) and *A New Theoretical and Practical Treatise on Navigation* (1836). On his return, he married Ann Herndon and was appointed to the navy's Depot of Charts and Instruments, headed by Charles Wilkes. Wilkes's major scientific interest was astronomy. He had an observatory in his house and was lobbying Congress to create the Naval Observatory. He was eventually successful.

Having achieved scientific fame on the basis of the book on navigation, Maury was appointed as the official astronomer on the U.S. Navy's voyage of exploration (1837), led by Charles Wilkes, but Maury did not go. He resigned after an argument with the secretary of the navy, Mahlon Dickerson. A serious stagecoach accident in 1839 left him lame and unfit for sea duty; he saw himself doomed to a life on shore. He worked on enlarging and improving his work on navigation and began a series of political papers urging naval reform.

In 1842, the Wilkes expedition returned; Wilkes's reward was the approval of the U.S. Naval Observatory. Maury was named head of the Depot of Charts and Instruments, and two years later, Naval Astronomer, a post he totally ignored. When the Naval Observatory moved out of Wilkes's house, a large number of ship's logs were found. Maury used these as raw data with which to chart circulation of the atmosphere and ocean currents, leading to the two-volume publication *Explanations and Sailing Directions to Accompany the Wind and Current Charts* in 1847. It became an outstanding navigational aid. By 1850, he was adding sailing directions that would decrease the time of an Atlantic crossing.

Years before, Faraday had shown oxygen to be paramagnetic; this means it can be aligned in the magnetic field of the Earth. Maury attempted to explain fluctuation in wind as a function of the Earth's magnetism. While his explanation was superseded by the better one arrived at by WILLIAM FERREL, it was a beginning in theoretical meteorology.

At his request, the U.S. Navy was taking soundings, using telegraphy, and dredging for biological samples. The samples retrieved were sent to the chemist and microscopist Jacob Bailey, at West Point, and to the biologist CHRISTIAN EHRENBERG, in Germany, for analysis. Telegraphy fascinated Maury, who envisioned its use in meteorology. At the time, he was engaged in a heated argument with Joseph Henry, the head of the Smithsonian Institution, and Henry's friend

Matthew Fontaine Maury, American seaman and head of Depot of Charts and Instruments and the naval observatory, 1842–61. *(Hulton/Archive/Getty Images)*

ALEXANDER BACHE. All three men were founding members of the American Association for the Advancement of Science (1849), but when that organization formed a commission to explore meteorology on land, Maury was excluded. The first international meteorological conference was organized by Maury and held in Brussels in 1853, its subject matter restricted to oceanic meteorology. Maury's most important publication, *The Physical Geography of the Sea, and its Meteorology*, was published in 1855 and in England in 1860. The book was a resounding success and saw a long life in many reprintings and new editions.

When the Civil War began in 1861, Maury resigned from the U.S. Navy to become a commander in the Confederate Navy: he was sent to England to lobby for the Confederacy. After the war, he was commissioner for immigration in Maximilian's ill-fated Mexican Empire, returning to England when that failed. He supported himself by writing textbooks until he was allowed to return to the United States in 1868. The last years of his life were spent teaching meteorology at the Virginia Military Institute. He died in Lexington, Virginia, on February 1, 1873.

The title "father of oceanography" has been one awarded to Maury by later physical oceanographers. While that title can be applicable to several people, he did make considerable contributions to navigation, to meteorology, and to bathymetry. Another of his very significant publications was *Bathymetrical Map of the North Atlantic Basin* in 1854. This depiction of the sea bottom was the first careful mapping of the North Atlantic segment of the globe-encircling mid-ocean ridge system.

⊠   McCammon, Helen Mary
(1933–   )
American
Environmental Scientist

A native of Winnipeg, Manitoba, in Canada, Helen McCammon was born on August 16, 1933. She received her undergraduate training in geology at the University of Manitoba, earning a B.Sc. in 1955, and then moved with her husband, Richard, to the United States. She continued her training at the University of Michigan, receiving an M.S. in 1956, and a Ph.D. in geology from Indiana University in 1959. Beginning as an undergraduate, McCammon worked for the Manitoba Bureau of Mines from 1952 to 1959. After some time as a field geologist in North Dakota, McCammon found her research interests moving from geology to environmental studies. She began studying marine physiology and ecology to better understand the effect of oil drilling and spillage on the environment. This continues to be her major research interest.

Working around family responsibilities, McCammon was an associate professor of geology at

the University of Pittsburgh (1963–68); then a visiting professor at the University of Illinois in Chicago, while working as a research associate at the Field Museum there (1968–70). She published several papers on brachiopods and then moved to the Environmental Protection Agency. McCammon was a researcher there from 1969 to 1972 and then became a member of the permanent staff of the U.S. Department of Energy where she is the director of the Environmental Science Division. She is a member of the American Society of Limnology and Oceanography, the American Association for the Advancement of Science, the American Geological Institute, and the Oceanographic Society.

⊠   Medwin, Herman
(1920–   )
American
Physicist

Herman Medwin, born on April 9, 1920, is a native of Springfield, Massachusetts. After graduating from the Worcester (Massachusetts) Polytechnic Institute in 1941, he spent the rest of his professional career in the western United States. He received an M.S. in geology from the University of California at Los Angeles (UCLA) in 1948, and a Ph.D. in physics there in 1954. This was an exciting time for physicists. Medwin taught physics at UCLA and at the Los Angeles City College while pursuing research on acoustics.

Medwin spent the 1961–62 academic year at the Office of Naval Research working on the transmission of sound in the sea. He served as a visiting scientist in Great Britain, Japan, and China. He then became chairman of the Acoustical Oceanography Group in Woods Hole, Massachusetts, before returning to California. He is now emeritus professor of physics at the Naval Postgraduate School in Monterey, California, and founder and chief executive officer of Ocean Acoustics Associates in Pebble Beach, California.

The latter is a consulting group that works on a variety of structures where sound is a factor. These range from underwater sounds to highways and tunnels to buildings and concert halls.

The focus of Medwin's research is sound scatter in the ocean and on its surface, high intensity sound and its effects, acoustic diffraction, and the effects of bubbles on sound. All of these are factors in determining the direction and velocity of sound in water and on it. The methods of acoustical oceanography depend on the spatial and temporal variability of the ocean. In the past 20 to 30 years, passive and active ocean acoustic characteristics have been "inverted" to take account of the physical and biological sources of background sounds, ranging from rainfall to whales. Some of the tools for acoustical inversions illustrate careful experiments at sea and/or in the laboratory, as well as mathematical-physical descriptions of wave phenomena, such as speed of propagation, dispersion, refraction, facet reflection, wedge diffraction, absorption, Doppler effect, and body resonance.

Medwin's books, *Acoustical Oceanography: Principles and Applications* (1977) and *Fundamentals of Acoustical Oceanography* (1997) were both coauthored by Clarence Clay. They are the standard works in this field.

Henry Menard *(U.S. Geological Survey, courtesy American Geophysical Union and AIP Emilio Segrè Visual Archives)*

⊠   **Menard, Henry William**
(1920–1986)
American
Geologist, Oceanographer

Henry Menard, a native Californian, was born in Fresno on December 10, 1920, and attended local schools for his early academic career, earning a B.S. and M.S. in geology from the California Institute of Technology in 1942 and 1947 respectively. Between the two degrees he served in the South Pacific during World War II. He completed work for his Ph.D. in marine geology at Harvard in 1949.

Menard's first professional appointment after receiving a Ph.D. was in the Sea Floor Studies Section of the U.S. Navy's Electronics Laboratory in San Diego. He then joined the Scripps Institution of Oceanography, La Jolla, California (SIO) in 1955 with the rank of assistant professor. In that position, he was involved in 20 oceangoing expeditions and more than 1,000 free dives using aqualung equipment. He was interested in this equipment and, in the 1950s, operated a scuba diving business. In 1961, Menard was appointed professor at the University of California at San Diego but left after a year when he became a technical adviser in the Office of Science and Technology in Washington, D.C. (1965–66). He returned to the University of California's Scripps Institution of Marine Resources. Menard wrote several books that help to teach marine geology, including his textbook,

Marine Geology of the Pacific, written in 1964 between appointments. *Anatomy of an Expedition*, published in 1969, was an account of one of his oceangoing expeditions. Menard was elected to the National Academy of Sciences in 1968.

His textbook *Geology, Resources, and Society*, published in 1974, discusses among other things the exploration for oil and factors affecting that. His earlier book, *Science: Growth and Change*, was published in 1971. It was a technical account of his work for the Office of Science and Technology, and was well received by historians of science, but not a popular success.

In 1978, Menard gave up teaching and served as director of the U.S. Coast and Geodetic Survey. In this position, he traveled extensively in the United States to various university geology departments and laboratories, and to branches of the Geodetic Survey. His opportunities for foreign travel were also extensive. Menard left this exciting position in 1981 and returned to the Scripps Institution, where he continued to research, teach, and write. He died in La Jolla, California, on February 9, 1986.

Two of his books were published in that year: *Islands*, a popular work on island formation, and *Ocean of Truth*, a memoir of his involvement in the study of global tectonics. Menard was an early proponent of the theory of plate tectonics that ultimately became recognized as the way the Earth creates new rock and destroys the old. This theory is based on the concept of rigid plates of Earth's crust—some of them bearing the continents—floating on the layer of magma below. His papers are part of the archives of the Scripps Institution.

⊠   **Mesyatsev, Ivan Illarionivich**
(1885–1940)
Russian
Oceanographer

Nothing is known of this oceanographer before his appearance in 1908 as an undergraduate in the department of mathematics and physics of the University of Moscow. He received a B.S. degree in 1912 and remained at the university in the department of invertebrate zoology, working on an advanced degree in that field. During the upheavals of World War I and the revolution, he apparently remained at the university, and by 1920 was the head of a group planning the first oceanographic institute in the Soviet Union. The institute, named Plovmornin, was established in 1921; its specific charge was to investigate all aspects of the Barents Sea. Starting in 1927, Mesyatsev was the director of the State Oceanographic Institute and by the next year, he was director of the Plovmornin as well. He maintained these posts for the rest of his life.

Mesyatsev's primary interest was the biology and environment of the Arctic Ocean. Many of his Soviet contemporaries were primarily interested in the resources of the region, but saw them only in terms of exploitation. Therefore, much of Mesyatsev's research, and all of the funds the government provided for its continuance, had to be concentrated on the movement, growth, and health of the fish populations in Arctic waters. His work pointed to the large, albeit fragile populations and ecosystems in polar waters. Much of his information was useful in later years, when Arctic oil spills necessitated extensive cleaning operations. Mesyatsev died in Moscow on August 29, 1940.

⊠   **Milne-Edwards, Henri**
(1800–1885)
French
Marine Biologist

Henri Milne-Edwards was born in Bruges (now in Belgium, then part of France) on October 23, 1800. His father, William Edwards, had retired there with his second wife, Elizabeth Vaux, after making his fortune on his West Indian plantation. Henri Edwards was orphaned young and raised by his half-sister and her husband, a Mr. Milne. On reaching the age of 21, Henri adopted his new surname, Milne-Edwards, and French citizenship.

The start of his professional career was his move to Paris, where he studied zoology while serving as professor of hygiene and natural history at *l'École centrale des arts et manufactures*. This school might be compared to a modern vocational high school. Milne-Edwards's main research interest was invertebrates—in his time, an unusual course of study. His achievements in this area were recognized by the scientific community. He was awarded the Academy of Sciences prize in experimental physiology in 1828, and 10 years later was elected to the Zoology Section of the academy; this was a teaching position in the Faculty of Sciences. The first works published by Milne-Edwards were those written with JEAN AUDOUIN, a fellow physiologist and friend. The two were involved in the morphology and physiology of crustaceans, and published together until Audouin died in 1841. Afterward Milne-Edwards continued their efforts alone. *Histoire naturelle des crustacés* (Natural history of crustaceans) was published in installments between 1834 and 1840. The next large publication, *Histoire naturelle de coralliares* (Natural history of corals), appeared in installments from 1858 to 1860. His last major work on specific organisms was also published serially, beginning in 1868 and finishing in 1874, *Monographie des polyps de terrains paléozoique* (Monograph on the paleozoic polyps).

At the end of his active teaching career, Milne-Edwards published his collected lectures in 14 volumes. In this massive work he concentrated on the physiology and comparative anatomy of humans and other animals. He conducted some of the earliest work on cell differentiation and the specialization of groups of cells that leads to the formation of tissue and, ultimately, organs. After his death in Paris on July 29, 1885, he was much honored by the scientific community and his many *anciens élèves* (old students).

⊠ Möbius, Karl August
(1825–1908)
German
Zoologist

Karl August was born on February 7, 1825, in Eilenburg, then in Saxony, the son of a wheelwright, Gotlob Möbius, and his wife, Sophie. He was lucky to have been educated beyond the local school. In the normal course of events, he would have entered a trade, most likely that of his father. Instead, Möbius trained as an elementary school teacher and taught in Harz mountain village schools from 1844 to 1849. In pursuit of his growing interest in natural science, he sat for a placement examination at the University of Berlin. Accepted, he studied under JOHANNES MULLER and CHRISTIAN EHRENBERG, who were his mentors. Möbius had hoped to become a part of a major scientific expedition, but when that did not happen he returned to teaching, this time in Hamburg. Once there, he joined the administration of the Hamburg Museum of Natural History, and in 1863 was cofounder of the zoo. This zoo included the first aquarium in Germany.

Möbius's career progressed quickly after the public took notice of the zoo and aquarium. By 1868, he was chairman of the zoology department at the University of Kiel and in the next year commissioned by the Prussian government to investigate mussel cultivation. The government was interested in growing mussels as a cash crop, as the British and French had been doing for years. This task led to extensive research on the North Sea and Baltic coasts. His wish for an expedition was granted, and in 1874 he sailed to Mauritius and the Seychelles to examine the fauna and coral reefs of these Indian Ocean islands.

The work on the tropical islands contrasted with the studies of northern European waters, and led him to develop what he called *Bicönose*, a concept that might now be translated as ecosystem. Möbius examined animals in their life communities and the relationship of place to organism.

In recognition of his growing fame, the university built a new Zoological Institute in 1881, designed according to Möbius's specifications. When Berlin built a new Natural History Museum in 1887, Möbius was invited to become its director. He accepted and served in this position until his death on April 26, 1908.

Möbius's significant work was marine biology and applied research growing from it. His major publication was *Die Auster* (The oyster), a detailed description of oyster fisheries, pearl formation, and cultivation. He also worked extensively on whale anatomy and on the foraminifera.

⊠ Mohorovičić, Andrija
(1857–1936)
Croatian
Meteorologist, Seismologist

Born on January 23, 1857, in Vorlosko, Croatia— then part of the Austro-Hungarian empire—Andrija Mohorovičić went to gymnasium (the high school equivalent) in Rijeka and then to the faculty of philosophy at the University of Prague. He graduated with a bachelor's degree in mathematics and physics in 1875. Returning home, he taught in elementary and then high schools in Zagreb and Osijek, then transferred in 1882 to the Nautical School in Bakar on the Dalmatian coast, where he established a weather station. In 1891, he moved to the main technical high school in Zagreb, and the next year he became head of the Meteorological Observatory in Zagreb. The University of Zagreb awarded Mohorovičić a doctorate in 1897 and the position of lecturer. He was promoted to reader, an academic rank equivalent to associate professor, in 1910.

When Mohorovičić developed the weather station, the official weather service of the Austro-Hungarian Empire, of which Croatia was then a part, was centered in Budapest. After a long battle, he was able to decentralize this service in 1900 and make the Zagreb Observatory the official one for Croatia and Slavonia. Research efforts there focused on tracking storms and seeking ways to utilize wind power as an energy source. His last paper in meteorology, discussing the effect of elevation on temperatures, was published in 1901. Then a new interest captivated him.

Seismology was a growing area of research, and the earthquake in the Kupa Valley on October 8, 1909, provided Mohorovičić with ample data with which to pursue his seismic studies. Mohorovičić detailed the transmission of seismic waves through the Earth. He found two sets of seismic waves, primary (P) and secondary (S); one pair moves much more slowly than the other. He correctly reasoned that the focus of an earthquake was within the outer layer of the Earth. He reasoned that the slower wave pair traveled to the observing station through that outer layer, and that the faster waves went down below the layer where the epicenter was, to a boundary layer, and were then bounced back to the detectors on the Earth's surface. That boundary layer is known as the Mohorovičić discontinuity or, more simply, the "Moho layer."

Calculating the time traveled by the P and S waves, Mohorovičić calculated the thickness of the outer layer of Earth to be about 50–58 km (30–35 miles). This was found to be true worldwide. Further researchers found that this outermost layer, the crust, had an average thickness of about 32 km (20 miles) on land, but only 5–10 km (3–5 miles) under the ocean. The separation between the crust and the mantle below it is the Moho layer. In 1957, the International Association of Seismology and Physics proposed to drill down almost to the Mohorovičić discontinuity. In the popular press this became known as the "Mohole," and then the "Mohole project"; it proved that Mohorovičić's theory of a boundary was correct, that this boundary level was worldwide, and that it was much closer to the surface of Earth in the oceans. It was an integral part of the experiments that verified the theory of plate tectonics.

Mohorovičić remained active, exploring the possibility of earthquake-proof buildings and redesigning seismological instruments, many of which were never built because of chronic under-funding. He organized the former Royal Regional Center as the Geophysical Institute in 1921. He continued to work there until 1926 and died in Zagreb on December 18, 1936.

Moro, Antonio Lazzaro
(1687–1764)
Italian
Geologist, Paleontologist

This early geologist was born on March 16, 1687, in a small town near Friuli (then called Feltre) on the Dalmatian coast when that was part of Venice. He was ordained a priest probably before 1710, and was a professor of philosophy and director of the seminary in Feltre. By 1721, Moro had an international reputation as an expert identifier of fossils. In retirement, he continued to serve the church as chapel master and to work on his fossil collection. In 1741, a major work, *Dei crostacei e degli altri corpi marine che se trovano siu monti* (On crustaceans and several other marine bodies which one may find in mountains), provided a link with the works of LUIGI MARSIGLI and others. It provided a stratigraphic listing of plant and animal fossils that predated the works of William Smith in England and GEORGES CUVIER in France. Fossils are found in layers of sedimentary rock; particular fossils characterize the strata in which they are found.

Unlike the neptunist geologists, who believed that the whole Earth had originally been under water, Moro was more of a plutonist. These geologists theorized that volcanic activity caused islands and continents, and Moro agreed with this hypothesis. He was familiar with the accounts of ancient writers describing the appearance of new volcanic islands, such as Mea Kaumen, an island near Thera (also referred to by its Italian name, Santorini). Moro reasoned that the marine fossils on mountains are evidence of their rise because of volcanic activity, and that the deformation of rocks is the result of heat and pressure, which produce the gneiss and marble that one can see. In Moro's version of Earth's history, all living flora and fauna are habitat-dependent and evolved from earlier forms. There was violent opposition to Moro's inclusion of humans in this gradual change from earlier forms to present ones.

Moro's work was not just a bridge between Marsigli and later geologists. He was one of the first to expound the idea of the evolution of living things and the rocks themselves. If limestone can become marble, he reasoned, then just about anything on Earth can be affected by change in condition. This was a revolutionary idea. Moro's work was later studied by other plutonists, particularly James Hutton, and later formed a basic work for CHARLES LYELL.

Moro spent the rest of his long life corresponding and trading specimens with other geologists both in Italy and abroad. He died on April 12, 1764, in San Vito del Friuli, not far from his birthplace.

Morse, Edward Sylvester
(1838–1925)
American
Paleontologist, Zoologist

Morse was a ninth-generation New Englander. He was born in Portland, Maine, on June 18, 1838, into the family of Jonathan Morse, a staunch Calvinist and a deacon in his local church. As a child, Morse was an inattentive student but an avid collector of shells and insects. Thanks to an elder brother's efforts, at 16, he was working as a draftsman for a local company that made steam engines. Morse began a short but significant career at Bethel Academy in Bethel, Maine, in June 1856. There, Dr. Nathaniel True recognized the boy's brilliance and allowed him to work with visiting scientists who were studying the biology and mineralogy of Maine. While doing so, Morse discovered a small land snail in September of that year. The snail, *Tympanis morsei*, began his scientific career. In 1859, the Boston Society of Natural History declared Morse's snail a true and new species.

On May 27, 1859, Morse traveled to Cambridge, Massachusetts, to meet the famous professor, LOUIS AGASSIZ, then at the Lawrence Scientific School. Agassiz was undoubtedly the best-known zoologist in America and he took Morse on as a student with a specialty in conchology (study of shells). While working for Agassiz, Morse attempted to join a Maine Infantry regiment (1862) but was rejected; his tonsils were chronically infected. Remaining in Cambridge, he maintained a correspondence with John Mead Gould. The letters from this lifelong friend and childhood collecting partner are a significant body of information about the Civil War. Morse married Ellen Owen in Portland, Maine, in 1863, and continued his studies in Cambridge. Upon graduation he was appointed to the Essex Institute in Salem, Massachusetts, and in 1867 to the Peabody Academy of Science there, where he was curator of Radiata and Mollusca. The two institutions are now the Peabody Essex Museum.

Brachiopods were Morse's research interest at the Peabody. These bivalves are alive now and have a very long history in the fossil record. They superficially resemble clams. While at the Peabody, he was one of the founders of the periodical *The American Naturalist* in 1867. The next year he was appointed a fellow of the American Academy of Arts and Sciences. This association elected him vice president in 1879 and president in 1886. Morse was professor of zoology and comparative anatomy at Bowdoin College in Maine from 1871 to 1874, when Harvard appointed him as a lecturer.

After publishing *A First Book of Zoology* in 1877, and while continuing the study of brachiopods, Morse accepted a position as professor at Tokyo's Imperial University. This part of his career lasted from 1877 to 1880. While in Japan, Morse began seriously collecting the local ce-

ramics. Both the Japanese collection and his teaching career produced several results: all brachiopods were moved out of the Mollusca and into their own class in the phylum Lophophorata; a marvelous collection of Japanese pottery and other artifacts was brought to America, and Morse's experiences, along with those of William E. Griffis and Lafcadio Hearn, formed the basis of the book on Japanese culture, *Mirror in the Shrine*. Upon his return to Salem, Morse resumed his directorship of the Peabody, a position he held until his retirement and emeritus status in 1916.

Bowdoin College awarded Morse an honorary Ph.D. in 1871. The scientific community honored him with appointment as Fellow in the National Academy of Science in 1876 and a D.Sc. from Yale in 1918. In 1898, he was the first American to receive the Order of the Rising Sun, third class, from the Japanese empire, which also gave him the Order of the Sacred Treasure, second class, in 1922.

Morse's lifelong interest in art was expressed in several ways. He illustrated his own work and that of others, notably a book by his friend John Mead Gould. After his retirement from Harvard, he embarked on a second career as curator of Oriental art and then museum director. He was president of the American Association of Museums in 1911.

Morse died at home in Salem on December 20, 1925, never having lost his sense of curiosity or his sense of humor, willing his brain to the Wister Institute in Philadelphia. He wanted that institution to study ambidexterity.

⊠   **Müller, Johannes Peter**
(1801–1858)
German
Comparative Anatomist, Physiologist

Johannes Müller, a shoemaker's son, was born in Koblenz on July 1, 1801, and spent his youth in the Moselle Valley. In 1819, he entered the university in Bonn and in 1822 was awarded a degree in medicine. Since he really did not want to be a doctor, he then went to Berlin to work with Carl Rudolphe, the noted anatomist. Rudolphe was an early scientist who was seriously engaged in removing the Goethe-inspired Romantic and philosophical baggage from experimental science. Müller worked with Rudolphe for two years and when, in 1824, he passed the state licensing examination in medicine, he returned to Bonn as a lecturer in physiology and comparative anatomy. The next year, pathology was added to his duties and by 1830 he was full professor.

His early work in neurophysiology was brilliant. In the 1820s, he worked on the distinction between motor and sensory nerves, the cranial nerves, and the nature of sensation. This work continued until about 1840, when he published his definitive two-volume work, *Handbuch der Physiologie des Menschen* (Handbook of human physiology). Müller had always hoped to succeed Rudolphe in Berlin. After turning down several offers, when Rudolphe retired, Müller wrote to the Prussian minister of education describing the requirements for the position of professor of anatomy in Berlin's Faculty of Medicine and outlining his own qualifications. He got the job.

Once in Berlin, Müller did no more experimental physiology but concentrated on comparative anatomy and the zoology of marine organisms. These he collected on almost annual field trips to the North Sea and Mediterranean coasts. In 1832, he published the definitive classification of amphibians and reptiles. This work was followed in 1834 by the study of Cyclostomata, an organism that Müller recognized as a fish and a relative of sharks. He also reaffirmed the Aristotelian observations—which had been repeated by Rondelet in 1555 and then by Sten in 1673—that some sharks, such as dogfish, are oviviparous: they produce live young.

Müller was a member of the Prussian Academy of Sciences by 1834, a recipient of the Copley Medal from the Royal Society, and the Cuvier Award of the Academy of Sciences in Paris. He served as rector of the university in Berlin for the 1838–39 year, and again for 1847–48. During the latter year, with its attendant revolutionary upheavals, Müller began to experience serious bouts of depression. He continued his painstaking work on echinoderms and sea slugs, carefully studying the complex maturation processes of the former. He maintained that physiology as expressed by physics and chemistry could not answer the question, What is life? but neither could the philosophical legacy of Goethe. Müller believed in a "vital force" as the answer to the classic question. He felt crowded out of science by the physicists and chemists who relegated comparative anatomy to a second-class status.

Having left specific instructions forbidding an autopsy, Müller died on April 28, 1858, in his Berlin home, a probable suicide. His work on the radiolaria and foraminifera was continued by ERNST HAECKEL, one of his many famous students.

Munk, Walter Heinrich
(1917–   )
Austrian/American
Geophysicist

Walter Munk was born into a wealthy, cosmopolitan Viennese family on October 19, 1917. His parents, Hans and Rega (Brunner) Munk, divorced, and his mother married Rudolf Engelsberg. When he was 14, the family sent him to school in New York because it was expected that he would be trained in banking; it was his maternal grandfather's business. He was expected to work in the New York offices of the family's banking business. He did that for three years while in high school and discovered that he hated banking. Enrolling as a physics major in the California Institute of Technology (CIT), he eventually earned a B.S. in 1939. That summer, Munk applied for a temporary job at the Scripps Institution of Oceanography, California, and met HARALD SVERDRUP, then looking for work as well. Sverdrup accepted Munk as a doctoral student before Munk finished his work on a master's degree in physics from CIT. That work was completed in 1940. Munk applied for American citizenship after the *Anschluss,* when Germany and Austria became one nation in 1938. His parents left Austria shortly thereafter, settling in Pasadena.

When World War II began, Munk enlisted as a private, hoping for a posting in the ski troops, but he was excused and assigned a research project. He and Sverdrup were to develop a methodology to predict surf on the coast of northwestern Africa where an invasion was planned. The area has predictable heavy surf in winter, and their goal was to find the appropriate weather conditions that would minimize landing difficulties. The two researchers, together with JAKOB BJERKNES and Jorgen Holmboe, taught and trained military meteorologists who eventually worked on the coasts of Africa, the Pacific Islands, and Normandy.

The University of California at Los Angeles granted Munk a Ph.D. in physics in 1947 for work done at Scripps. That summer saw him aboard a navy vessel studying the operation of submarines in Arctic waters. This was followed by six months at the University of Oslo (1949) studying ocean currents on a Guggenheim fellowship.

The postwar years brought a large investment in basic science by the both the military and the National Science Foundation. This major research effort included radar improvements, acoustics, seismology, and satellite tracking and funded basic studies on global phenomena such as oceanic currents, circulation, tides, and wave action. By training, inclination, and location at Scripps, Munk was part of this research effort and produced many papers and reports in a number of areas. The international efforts of exploration in the 1950s afforded Munk the opportunity to broaden his research interests. He participated as one of the scuba divers in the 1952–53 Capricorn Expedition, a scientific endeavor that explored Pacific deeps. He researched the Earth's wobble in its orbit and the resulting effect on gravity, studied deep ocean tides and their effect on oceanography, the origin of tsunamis, and the effects of wind on oceanic circulation.

Munk served on a committee of the International Geophysical Year (IGY), 1957–58. This work was exciting, but when it ended, Munk believed that Earth sciences were receiving less money and less public emphasis than space science. Together with HARRY HESS, Munk lobbied in the early 1960s for support for Project Mohole. This work would drill through the layers of Earth's crust to the discontinuity discovered by ANDRIJA MOHOROVIČIĆ, the boundary between the crust of the Earth and the mantle below it. Munk was involved in the preliminary work that proved that this work would be possible. Although controversial, the project drew international participation and opened a series of new fields, leading to a scientific revolution in the knowledge of geology and the history of Earth. The Mohole was expected to prove the plate tectonic theory. To its critics, it was an expensive, unnecessary exercise.

Munk continued to serve on scientific advisory committees; later in the 1960s, he was a member of a group advising the military, and served on several panels of the president's Science Advisory Committee. He continued research and traveled widely, generally while engaged in some scientific inquiry. His entire career was based at the Scripps Institution of Oceanography (SIO), a part of the University of California at San Diego. While Munk served on numerous faculty committees and the Faculty Senate, he refused the directorship of Scripps because the position would have reduced the time he could devote to research. Munk founded the Institute of Geophysics and Planetary Physics, a La Jolla branch of an institute at UCLA, believing, as did Sverdrup, that all the Earth sciences were interconnected. Between 1965 and 1975 he worked on the development of instruments for MODE (Mid-Ocean Dynamics Experiments) and studied internal waves in the ocean, which are the result of movements of water masses that differ in temperature and density. In the 1980s, he worked on acoustic tomography, a phenomenon he described as underwater weather. Sound travels differently in bodies of water that have differing properties. This phenomenon is measurable and yields much data on aquatic environments and the animals that inhabit them.

The Navy Secretary named Munk to one of the Secretary's Research Chairs in Oceanography in 1984. This was in recognition for outstanding work in the field. Walter Munk has not retired, and his work frequently appears in such publications as *Science, Journal of Climate,* and *Journal of the Acoustical Society of America.* He is working on tidal friction as a generator of internal waves, and on the seasonal variations in sea level. Acoustic tomography remains an interest.

⊠ Murray, George Robert Milne
(1858–1911)
British
Botanist

George Murray, born on November 11, 1858, in Arbroath, Scotland, was one of the eight children of George and Helen Murray. Education was provided to the children in the local schools. When Murray left school in 1875, he traveled to spend a year at the University of Strasbourg. On his return to Britain, he was appointed as an assistant in the botany department of the British Museum. This was his department for the rest of his working career, and he rose to the position of keeper (curator) in 1895. An expert in mycology (fungi) and taxonomy, Murray wrote a noted paper on fungi in 1878. On the basis of this important work, Murray was elected a Fellow of the Linnaean Society the same year. He wrote all the articles on fungi in the 1879 edition of *Encyclopaedia Britannica.* He was personally involved in the creation of the botany department of the British Museum and its arrangement when the science collections moved to their present location in South Kensington.

Murray's summers were reserved for collecting. A loyal Scot, Murray concentrated his algae and diatom collecting efforts on Scotland's lochs and surrounding seas. He used a floating laboratory while at sea, the *Garland,* a vessel owned by the Fisheries Board. Among his innovations were new collecting nets of fine-mesh silk, with which he trapped phytoplankton (very small, floating plant species). His most significant works on spore formation in diatoms were both published in 1897, *On Reproduction in Some Diatoms,* and *On the Nature of Coccospheres and Rhabdospheres.*

Going farther from Britain, Murray was the naturalist on an 1886 expedition to the West Indies financed and sponsored by the British Museum. The major objective of this trip was the observation of a solar eclipse. However, using his silk nets, Murray also collected and analyzed specimens. The taxonomic analysis of this work was not completed until later, appearing in installments between 1891 and 1897. That year, Murray returned to the West Indies to collect diatoms, particularly *Cocosphora,* a diatom with an unusual shell. Another ocean voyage in 1898, with Murray aboard as the plant expert, was sponsored by the Royal Geographic Society, to study the western Irish coast and collect plankton at the edge of the deep sea. On Murray's last oceanic voyage in 1901, he was the scientific director of the *Discovery* expedition to the Antarctic. The scientific procedures used were published as the *Antarctic Manual,* which Murray edited. This book was a major reference for scientific work in polar waters. Ill health prompted Murray to retire early, and he left the museum. He returned to Scotland in the spring of 1911 and lived in Stonehaven until his death on December 16, 1911.

⊠　**Murray, Sir John**
(1841–1914)
British
Oceanographer

John Murray was born in Coburg, Ontario, on March 3, 1841, to Robert and Elizabeth (Macfarlane) Murray, who had immigrated to Canada in 1834. The family went back to Scotland in 1858. The boy's education was determined by his grandfather, who sent him at first to secondary school in Stirling, Scotland, and then to the University of Edinburgh to study medicine. Oceanography was relegated to his spare time, but Murray did manage to sail on the *Jan Mayen* to Spitsbergen, Norway, in the Arctic Ocean, in 1868. His specific tasks on this voyage were to collect data on currents, temperatures, and movements of sea ice. His reports were sufficiently interesting to bring him to the attention of Peter Guthrie Tait, the physics professor at the University of Edinburgh, who gave him a job as physics laboratory assistant. Since he was not progressing toward his degree, and the laboratory position was not sufficient to support him, Murray left the university in 1872, without a degree, but with recommendations. Both Tait and Sir William Thomson, later Lord Kelvin, whom Murray had met by chance, wrote on his behalf to CHARLES WYVILLE THOMSON, who was Regius Professor of Natural History at the University of Edinburgh. Thomson was also the scientific director of an expedition being organized by the Royal Society for the Royal Navy. The result of this introduction was that Murray spent the next three years on board the HMS *Challenger*. There, he was the oldest, and in the very hierarchical society on board ship, the chief assistant. His specific charge was the mapping and classification of bottom sediments.

Before Murray's work, very little of the bottom sediment had been categorized. The habitats of the few calcareous (calcium-containing) remains found were unknown. They could have been surface dwellers or bottom dwellers or from any intermediate aquatic environment. Since so few animal remains were discovered, it was assumed that there was little life in the deeps; Murray classified and named the specimens found as diatom, radiolarian, and pteropod fragments. However, the greatest part of the bottom sediment was a rust-colored, finely divided material that contained continental dust and volcanic ash. Based on these findings, Murray concluded that animal remains in the ocean had either dissolved in the water or been covered up by the red clay.

When the *Challenger* returned in 1877, Murray remained with Thomson to organize the report, which was due in five years. Enlisting the aid of other naturalists, notably ALEXANDER AGASSIZ, Thomson and Murray divided the wealth of specimens brought back into categories, each of which was to be a separate monograph. The descriptive work had begun. Only the first of many volumes had appeared before Thomson's death in 1882. The British Treasury was the agency responsible for the work, and it appointed Murray to finish the report. When the time for producing the report approached, Murray realized the enormity of the remaining task and petitioned for another five years. The deadline was extended—the first of several extensions. The monographs were finally completed by 1895. Two of the last volumes printed were Murray's *Summary* in 1895, and the more important *Deep-Sea Deposits* (1891), which Murray wrote with ALPHONSE RENARD of Ghent as his coauthor. *Deep-Sea Deposits* was the first major work to treat the oceans as a single entity. It mapped areas and explained their unique characteristics in terms of temperature, flora, fauna, and sediment types. Murray was convinced that oceans had been a feature of Earth throughout geologic time.

In addition to the massive amount of work presented by the *Challenger Reports*—as individual monographs—Murray continued with his own research program. In 1880 he sailed on the *Knight*

Errant and two years later on the *Triton*. On the latter, Murray and coworkers located, named, and mapped the Wyville Thomson Ridge and explored its effect on the marine animal life near the Shetland Islands off the coast of Scotland. One theory proposed by Murray created much controversy; he speculated that coral reef development differed considerably from the process proposed by Darwin. Both ideas had their supporters, but Darwin's theory of coral accretion and oceanic subsidence that formed lagoons was conclusively proved in the mid-20th century. While studying coral islands, Murray forged a working relationship and then a friendship with Alexander Agassiz. Together, they found the manganese nodules that are a prominent feature of abyssal oceanic plains. Using deep sea sediment and rock samples, Murray in 1886 theorized the existence of a continental landmass under the Antarctic ice and began campaigning for another round of polar explorations. Other projects that interested him included the foundation of marine and meteorological stations. Murray did not forget he was a

Scot; he sat on Scotland's Fisheries Board and surveyed its freshwater lochs.

Murray financed much of his research himself, but his financial security was assured. He married Isabel Henderson in 1889. She was the daughter of a shipbuilder and brought her fortune with her. Murray made another fortune after he analyzed a Christmas Island rock that had been sent to him by a *Challenger* shipmate. It proved to be rich in phosphates, and after Great Britain annexed the island in 1887, Murray was given a leasehold in 1891. Mining of this material used in fertilizers began in 1900, and the profits supported the *Challenger* work and another expedition aboard the *Michael Sars* in 1910. The book based on this four-month trip was written with Johan Hjort; when it appeared in 1912, it immediately became the standard work on oceanography. An introductory work, *The Ocean*, was published the next year. Murray was killed in an automobile accident on March 16, 1914, in Kirkliston, Scotland. He bequeathed his fortune to further oceanographic research.

N

⊠ Nansen, Fridtjof
(1861–1930)
Norwegian
Biologist, Explorer

Fridtjof Nansen, a man of several careers, was born on October 10, 1861, in Frøen, a small town that was then near fields and forests outside of Christiania, Norway. The area is now part of metropolitan Oslo. His father was a lawyer; thus the family was reasonably well off, and the boy's childhood recreation took place largely outdoors, on his own in the woods. This gave him a strong sense of the physical world, and he entered the University of Christiania in 1880, to specialize in zoology. After graduation with a degree in that field, Nansen spent the spring and summer of 1882 aboard the *Viking,* a sealer bound for Jan Mayen Island. This began his fascination with the Arctic. Upon his return, Nansen was appointed curator at the Bergen Natural History Museum, which became his base for research. He is remembered in a piece of equipment, the Nansen bottle, a self-closing container for taking water samples, which has been modified and perfected since he first made it. He accepted the post of professor of zoology (1896) and then professor of oceanography (1908) at the Royal Frederick University in Christiania.

Nansen first researched the nervous system of invertebrates. He presented his thesis on that sub-ject in 1888 and was granted a doctorate by the University of Christiania. His love of the Arctic and intent to return there were continuing notes in his work. He made plans for an expedition into the interior of Greenland in 1887. After finishing the doctorate, Nansen and five companions set out to cross the ice cap from east to west on skis. Nansen had grown up using skis; his mother was an early enthusiast for this mode of travel and a pioneer in women's use of skis. The choice of direction was likewise simple. The six men would move from the wild east coast to the inhabited west, where they would spend the winter. Expert opinion had it that the six would perish, and they nearly did; they were delayed in starting and finally began their ascent of the glacier on August 15, late in the short Arctic summer. Despite some extraordinary hardships, they arrived at Godthaab in early October, having confirmed that interior Greenland is covered with an ice sheet. They also brought international attention to the use of skis. One result of the winter in Godthaab was Nansen's book on Inuit culture, *Eskimoliv* (Eskimo life), which was published in 1891.

The Greenland exploit was followed by a scientifically more important one, the expedition of the *Fram.* The ship's name means "Forward," and it was designed to withstand the pressure of the ever-present ice. Leaving in the summer of 1893, Nansen set out to prove that a westerly ocean current existed. He based this theory on the evidence of the *Jeanette*, an American ship that was wrecked off the New Siberian Islands and eventually drifted toward Greenland's southern coast. The *Fram* sailed east to the Siberian coast and was frozen into the ice as expected. Nansen hoped that the ice would carry the *Fram* close to the North Pole, but it did not. He set out with Hjalmar Johansen in March of 1895, equipped with skis, sleds, dogs, kayaks, and food for 100 days, with the intent of reaching the North Pole and then returning to some human settlement. When they left the *Fram*, it was at 84° north latitude. Nansen and Johansen got as far north as 86°14', the farthest point reached by anyone at that time and 320 km (200 miles) from the North Pole. They wintered on Franz Josef Land after a trek of 132 days. In the summer of 1896, they resumed their trek south and were picked up by Frederick Jackson, an English explorer who returned them to Norway in August 1896. The *Fram* returned at almost the same time, having drifted in the ice for three years emerging north of Spitsbergen.

The right-hand drift that moves air and water is most noticeable in the Arctic, driving all the water movement in the area, and creating the weather patterns all over the Northern Hemisphere. This fact has occupied the studies of many meteorologists and students of ocean currents. Nansen attempted to demonstrate the rate of movement and its predictability, and thereby establish the validity of the theoretical work done by WILLIAM FERREL and VAGN EKMAN. He became a popular hero, since his attempts to reach the North Pole and survive captured the nation's imagination. The news of the *Fram* went round the world. As a result, Nansen gained an unparalleled reputation as a polar explorer and organizer. The book he wrote detailing the *Fram* expedition was an international best-seller and won its author speaking engagements in many geographical societies.

Nansen became the most famous Norwegian of his age. He continued to serve his university and functioned in other areas as well. As the leading

oceanographer of his day, he was instrumental in fostering such projects as the establishment of a scientific research center in Bergen and the work of VILHELM BJERKNES and his son JAKOB BJERKNES. In 1905, Nansen played a role in the dissolution of the union of Norway with Sweden and then served the new Norwegian government as ambassador to Great Britain from 1906 to 1908. During the Great War (World War I), Nansen again served as a diplomat, negotiating with the United States after the American entry into the war and the resulting blockade of neutral Norway's trade routes. Because he was well known, any cause that he supported, either scientific or just heroic, became a national imperative. In 1917 Nansen persuaded Vilhelm Bjerknes to return to Norway from Leipzig, Germany. Not only did Bjerknes do so, but Nansen also persuaded the government to create a weather research station for him in Bergen.

The war totally changed Nansen. He was so appalled by the mass slaughter and displacement of people that he devoted the rest of his life to the prevention of war, the League of Nations that was supposed to accomplish this, an international judicial system, and the plight of refugees and prisoners. One cause led to another—famine in Russia, displacement of Greek minorities in Turkey and Turkish minorities in Greece. Nansen was involved in trying to help them all. In 1922 he was awarded the Nobel Prize for his organization of relief for Russian famine victims. He literally worked himself to death in humanitarian causes, dying on May 13, 1930. He was buried on May 17, Norway's national day.

⊠ Neumayr, Melchior
(1845–1890)
Austrian
Geologist, Paleontologist

Melchior Neumayr was born in Munich on October 24, 1845, the son of Max von Neumayr, a Bavarian government official. The boy was educated in Munich and then Stuttgart. He was sent to the University of Munich. Using his father's career as an example, he enrolled in the faculty of law. He quickly changed his mind and studied paleontology and geology. The University of Munich granted Neumayr a doctorate in geology in 1969. His first employment as a geologist was in Vienna where he served in the Imperial Austrian Geological Survey.

The post of *privat docent* in paleontology and stratigraphy at the University in Heidelberg was offered to Neumayr in 1872 and he accepted. This position has no regular equivalent in American university structure; it might be described as a tutorial post. Stratigraphy is the study of layers of sedimentary rock and the fossils found in it. Based on layering, one can assign an age and origin to the specimens found.

In 1879, after his marriage to Paula Suess, daughter of EDUARD SUESS, the professorship of paleontology at the University of Vienna was created for him. Suess was the professor of geology at the University of Vienna, a court figure, a popular lecturer, and a politician.

Through the 1870s, Neumayr traveled and collected in southern Germany, the eastern Alps, and the Carpathians. He became expert in the life forms of the Jurassic period and traced the development in animal shells into the Tertiary period. He had been a devoted student of the work of CHARLES DARWIN, who had developed a workable theory of evolution. He used Darwin's theories to demonstrate the evolution and radiation of earlier species into later ones. On several scientific trips to the Greek mainland and islands (1874–76), Neumayr collected both plant and animal specimens. Using these artifacts, he created a classification of geologic structures based on the fossils of the region. This was published, as *Zum Geshichte des östlichen Mittelmeerbecken* (On the geologic history of the eastern Mediterranean) in Berlin in 1882.

Following Suess's lead, Neumayr separated the ammonite fossils by geological time period, thus drawing clear lines of descent from Jurassic to Cretaceous to Tertiary periods. Ammonites were a numerous and formerly successful class of the phylum Mollusca. They are all extinct; their closest living relative is the chambered nautilus. Further work examined the relationship of climate to the fossil finds of the Jurassic and Cretaceous periods, "Uber Klimatische Zonen während der Kreide und Jurazeit" (On climatic zones during the Cretaceous and Jurassic times). This work was published in *Denkschriften der Akademie der Wissenschaften* (Scholarly writings of the Academy of Sciences) in 1883.

Although suffering from a heart ailment, Neumayr continued to work on his *Die Stämme de Thierreiches* (The positions of the animal kingdoms). This was dated 1889 but did not appear until after Neumayr's death, at home in Vienna, on January 29, 1890.

In this major work, he drew parallels between paleontology and zoology, proving that paleontology was much more than a classification table for index fossils. It is the outline for the pattern of descent and change that might now be called genetic drift.

Nordenskjöld, (Nils) Adolf Erik
(1832–1901)
Swedish
Geologist, Explorer

Adolf Erik Nordenskjöld, was born in Helsinki, Finland, on January 18, 1832. The Nordenskjöld family had supplied the Swedish government with a succession of soldiers, scientists, and royal administrators. During Adolf's university days, the administration of the University of Helsinki was controlled by Russian overseers. Nordenskjöld's outspoken criticism of them meant that after graduating in 1858 with a bachelor's degree in arts and letters—the only degree given then—he would have no future in Finland. However, by the time of his graduation he had achieved a reputation as a mineralogist and consequently was appointed chief of the Mineralogical Division of the National Museum in Stockholm. The city would remain his home base for the duration of his career.

Nordenskjöld's work as an explorer and geologist began in 1858 with the first of five field trips to Spitsbergen, an archipelago in the Arctic Ocean north of Norway. The northernmost point he reached was 81° N latitude.

Over the next 30 years, Nordenskjöld undertook 12 exploratory trips to the Arctic; the most significant was the voyage of the *Vega*. This journey in 1878–79 saw the first successful transit of the Northeast Passage, the polar waters north of Scandinavia and Siberia that stretch from the Atlantic to the Pacific Oceans. Nordenskjöld was responsible for introducing serious science to Arctic exploration, planning the scientific work, reviewing the equipment, and chosing the staff. The party left Tromsø, Norway, in July 1878, spent the winter trapped in the ice of the Bering Strait, and entered the Pacific the following July. Nordenskjöld's five-volume work, *Voyage of the Vega*, included material on geodesy, geomagnetism and geophysics, and ecological and anthropological studies of the regions traversed.

In addition to this major work, Nordenskjöld also produced two atlases: *Periplus* (1897) and *Facsimile Atlas* (1898). While maps can be distorted, flat representations of a curved surface, Nordenskjöld attempted to show the true size and shape of geographical structures in relation to their latitudes and longitudes. These map projections were a major influence in the history and development of maps and mapmaking.

Nordenskjöld was considered one of the most significant figures in natural science in all of Scandinavia. He died at his home in Dolbyö, Sweden on August 12, 1901.

P

Penck, Albrecht
(1858–1945)
German
Geologist

Albrecht Penck was born in Leipzig, Germany, on September 25, 1858. He attended the city's university, graduating in 1875 with a degree in geology. He taught and was rector for a short time in the University of Munich before his appointment as professor of physical geography at the University of Vienna in 1885. In 1906, Penck moved to Berlin, as professor of geography.

A major influence on Penck's work was a lecture delivered by Otto Torell in 1875. Torell spoke of glacial ice. This is not just floating water; it is a major element in shaping the land over which it moves. Penck's research on the Quaternary geologic period resulted in the publication of a major work coauthored by Eduard Bruckner, the three-volume *Die Alpen im Eiszeitalter* (The Alps in the ice ages), which appeared in installments between 1901 and 1909. In their work, the two geologists identified the four ice ages of the European Pleistocene epoch, Gund, Mindel, Riss, and Wurm, after the river valleys that were the first evidence of each ice age. The term *geomorphology* is believed to have been created by

Penck. It is used in his book on the subject, *Morphologie der Erdoberfläche* (Morphology of the Earth's surface), published in 1894.

The devastation of World War I (1914–18) greatly affected Penck, as did the tragic death of his son Walther Penck, also a geologist, in 1923. The new Institut fur Meerskunde (Oceanographic Institute), connected to the University in Berlin, engrossed him and gave him a base from which to sponsor the *Meteor* expedition of 1925–27. That scientific effort, led by August Merz, was charged with taking soundings in the South Atlantic.

As early as 1891, at the Fifth International Geographic Congress in Berne, Switzerland, Penck was advocating uniform reportage and spelling of names of Earth's features and landmasses. He continued his campaign throughout his working life and even after he retired in 1926. This sensible idea was gradually implemented in France, Germany, and Great Britain in the years before World War I. By 1931, fewer than 100 of the estimated 840 maps needed to map the whole Earth had been produced. This project was completed during the 1960s and 1970s, in part with the use of satellites. It is a fitting tribute to Penck, who died in Prague on March 7, 1945.

⊗ **Pettersson, Hans**
(1888–1966)
Swedish
Oceanographer

Hans Pettersson was born in Marstrand, Sweden, on August 26, 1888. His father was a professor of oceanography at the University of Uppsala, and Pettersson attended that university. He studied physics with Anders Ångstrom, the professor of physics and astronomy at the University of Uppsala, and then went to Great Britain to work on optics and radioactivity with Sir William Ramsay.

In 1913, Pettersson joined the staff of the Svenska Hydrografiska Biologika Commission (Swedish Hydrophysical Biological Commission) and worked on tides and currents in the Kattegat, the narrow waterway between Sweden and the Danish peninsula. This is an interesting body of water since it presents many areas of study; because the water in it is layered, there are boundary questions, as well as questions about water flow and densities. By 1914, Pettersson had a more permanent position in the Göteburg Högskola (Göteburg High School) as lecturer. He was at that post until 1930, and in addition to his teaching duties, continued his research on the interaction of ocean and climate as evinced by sea level changes.

Pettersson had been working with his father since his undergraduate days. Together they studied the heat capacity and movement of bodies of water within the North Atlantic. Other areas of interest and exploration were the radioactivity of seawater and bottom sediments and the measurement of insolation (amount of sunlight) at varying levels of water. Pettersson was an early expounder of the theory that undersea volcanic activity was a source of organic materials on the ocean floor and that the ocean floor was created by lava.

Years of planning, fundraising, and cajoling finally brought the Göteburg Oceanographic Institute into being in 1939. Pettersson served as its director from its opening until his retirement in 1956. While serving as director, he amassed private funds for a Swedish oceanographic venture on the *Albatross*. This expedition spanned several years 1946–48 and it was charged with sampling deep-sea sediments for radioactivity and conducting optical studies on varying layers of water. One later result of this work was studies of the ages of varying bodies of water using the decay of some radioactive elements as clocks.

Pettersson was awarded many honors in both Europe and the United States. On his retirement in 1956 he was elected a fellow of the Royal Society of London. He retired in 1956 and moved to Hawaii in the same year. He was professor of geophysics at the University of Hawaii for several years before his death in Göteburg, Sweden, on January 25, 1966.

⊗ **Piccard, Auguste**
(1884–1962)
Swiss
Physicist

Auguste Piccard and his twin brother, Jean-Felix, were born to Jules and Hélène (Haltenhoff) Piccard on January 28, 1884. They grew up in Basel, Switzerland, where their father was professor of chemistry at the university, part of a large, wealthy, and influential family. The twins' grandfather was a functionary in the federal Swiss government, and an uncle manufactured hydroelectric turbines. Both young men attended the Eidgenosse Technische Hochschule (Federal Technical University) in Zurich, Auguste specializing in mechanical engineering, Jean-Felix in chemical engineering. They then went on for doctoral work, Jean-Felix to Munich, where he worked with Adolph von Baeyer, then to Lausanne, the University of Chicago, Massachusetts Institute of Technology, and finally the University of Minnesota. He became a citizen of the United States in 1931. Auguste

Auguste Piccard (back row, first on the left) with one of the most impressive collections of science greats ever assembled. The photo was taken at the meeting of the Fifth Council of Physics, in Brussels, Belgium, October 1927. *(Photograph by Benjamin Couprie, Institut International de Physique Solvay, courtesy AIP Emilio Segrè Visual Archives)*

continued his studies in Zurich and later in Brussels, Belgium. He remained in Brussels from 1922 to 1954 as professor of physics at the Brussels Polytechnic University.

Auguste Piccard is a significant figure in marine science because of his diving vessels. He began exploring the limits of Earth with stratospheric balloon experiments. On May 27, 1931, Piccard and his coworker, Paul Kipfer, lifted off from an Augsburg, Germany, meadow and ascended to a record 477.8 m (15,775 ft). The air samples they collected were analyzed by Jean-Felix. After two more flights, Auguste Piccard turned his attention to Earth's other frontier, the sea. He envisioned a device that would do what the balloon had done, allow a man to go into an otherwise impossible environment.

World War II (1939–45) interrupted the work on the diving device that Piccard had named the bathyscaphe. The name was derived

from Greek roots for "deep boat." It was a hollow, metallic sphere designed to withstand water pressure of 843.6 kg/cm^2 (12,000 lbs/in^2). This was launched with considerable publicity in 1948 off the Cape Verde Islands. It did not work. A second attempt was made in 1953 after modifications. Piccard, with his son Jacques, dove down to 3,081 m (10,168 ft.). This is more than three times deeper than the record dive of CHARLES WILLIAM BEEBE, an earlier explorer of ocean deeps. Again working with his son, Piccard built another bathyscape, the *Trieste*, which he sold to the U.S. Navy. In 1960, Jacques Piccard, accompanied by Lt. Donald Walsh of the U.S. Navy, dove in it to a depth of 10,849 m (35,800 ft) in the Marianas Trench, the deepest part of the Pacific Ocean. Auguste Piccard died in his home in Lausanne, Switzerland, on March 24, 1962. He had been working on the design of yet another deep-sea diving chamber. His brother and collaborator had died on January 23, 1960, in Minneapolis, Minnesota. Both men were enthusiastic explorers and scientists who extended man's reach into hostile environments and collected data about them.

⊠   **Poli, Giuseppe Saverio**
(1746–1825)
Italian
Natural Scientist

Giuseppe Poli was born in Molfetta on October 28, 1746, son of Vitangelo and Eleanora, who were of the minor nobility. He attended the venerable university in Padua, where he studied in the Faculty of Medicine and attended lectures in natural science. Upon graduation, he worked as a physician in Molfetta and collected specimens, both plant and animal, of scientific interest. His hobbies were typical of a rich country gentleman. One was numismatics: he was a knowledgeable collector, considered an expert on ancient coins and medals. His scientific hobby, however, eventually became the driving force in his life.

To learn and collect more, Poli traveled in Italy to visit other universities, staying for months in each. He became known in scientific circles and was offered a chair in physics at the Royal Military Academy in Naples, capital city of the Kingdom of the Two Sicilies. This appointment sent Poli on an extensive instrument-buying trip. On his return he was offered the position of head of the physics department at the Hospital for the Incurables. This institution was in reality a research center, as well as a hospital. Ferdinand II appointed him tutor to his son, Francis, the heir to the throne. Poli tutored the prince, ran his physics departments, directed the military school, and was chair of the experimental physics program at the University of Naples.

His correspondence kept Poli in touch with scientists throughout Europe and he was a tire-less worker in the field of science appreciation. Since access to scientific materials was essential, he used his position in the court to persuade the royal family to open the royal library to scholars and to support a botanical garden, to which he donated his own plants. With the support of the court, Poli concentrated his marine studies on Mediterranean crustaceans and mollusks. He is said to be the founder of the study of the Mollusca, and the results of his efforts were published in two folio volumes, *Testacea utriusque Siciliae, 1781–1826*. These books are illustrated and hand-colored; they are very rare and expensive. Michele Troja was Poli's collaborator on this masterwork and it was he who finished it: the third volume appeared after Poli's death. The bulk of Poli's collection of marine specimens was willed to the Museo di storia naturale (Museum of Natural History).

In 1772, Poli's primary scientific interest was physics. He examined the electric phenomena and magnetism, concluding that lightning, electricity, and magnetism are all related. He was a century ahead of himself: all of these subjects were later reexamined by a number of physicists, and Poli's speculations were confirmed. Putting himself in harm's way as BENJAMIN FRANKLIN did on a later occasion, he remained on the deck of a ship en route from France to England during a thunderstorm to prove that lightning and electricity were the same. He did prove it; the ship was struck by lightning and almost lost. Unlike Franklin's similar experiment, Poli did not conduct the lightning to a safer place, the Leyden collecting container. The next natural phenomenon that was of interest to him was earthquakes. Poli spent the rest of his working life examining the effects of earthquakes and the anecdotal information of animals "sensing" a quake before it happened. He was a member of many academic societies both in Italy and abroad, a fellow of the Royal Society of London, and a much-admired member of the court. He died in Naples on April 7, 1825.

Other works written by Poli include the five-volume *Elementi di fisica sperimentale* (Elements of experimental physics), Naples, 1787, and *Memoria sul terromoto* (Memoirs of earthquakes), Naples, 1805.

⊠   **Portier, Paul**
     (1866–1962)
     French
     Biologist, Physiologist

Paul Portier was born in Bar-sur-Seine on May 22, 1866. His family was one that had always had several members involved in the bureaucratic affairs of the nation, and so when the young man passed the *bac* (the final examination after secondary school and the requisite for university study), he was offered a post in the Ministry of Finance. He refused and professed an interest in science. His family insisted that, if he chose a career outside the government, it then must be in medicine. Yielding on that point, Portier attended the medical faculty at the University of Paris and was awarded both an M.D. and a *docteur en science* (doctorate in sciences). He stayed on in the physiology department of the medical faculty.

The work that Portier is best known for was done in collaboration with CHARLES RICHET in 1901. They worked together on ALBERT I's research ship, the *Alice II*. Albert I, prince of Monaco, was a patron of scientists. He outfitted and sailed ships that were staffed by scientists who worked on a number of projects of interest to Prince Albert. The director of research on the ship was interested in *Physalia*, the Portuguese man-of-war, a colonial jellyfish that produces toxin potentially fatal to humans. Portier and Richet ground up the jellyfish and injected guinea pigs and pigeons with the toxin, which produced very deep sleep; more toxin killed them. The researchers tried to induce immunity using the same techniques used in research to prevent tetanus, anthrax, and diphtheria. They tried to attenuate (weaken) the toxin enough that a victim would not develop the illness but would be rendered immune. This work was too complex for a shipboard laboratory, and in 1902, the group brought their materials back to the Paris laboratories.

In their second round of experiments, they used the toxin produced by the sea anemone *Actinia* on dogs. The researchers were astonished to find that one dose of toxin—attenuated or not—produced a reaction, and another exposure produced anaphylaxis, a shock reaction that disrupts mechanisms controlling breathing and circulation, leading to death. They quickly realized that they had found not immunity but increased sensitization; they had discovered allergies.

Portier continued as a researcher in the physiology department. He studied the mechanics of insects "walking" on water, and in 1909, he published an analysis of the effect of surface tension on species of insects. Again examining the properties of water and how it affects life in the sea, he worked on the "blow" of whales and other marine mammals: Why is the water vapor that a whale blows into the air visible when the whale breaks through the surface? His work produced an answer: the water vapor in the exhaled air condenses because it is warmer than the comparatively cooler open air and no longer under pressure as it was inside the animal. Water in the exhalate then condenses, instantly forming visible droplets. This work had later ramifications in the design of self-contained underwater breathing equipment.

Working on another pressure problem, Portier examined the internal environment of fish as affected by the pressure and salinity of the waters surrounding them. How do fish manage to swim at varying depths without being crushed by increased water pressure or exploding if they rise to a higher level in the water? This was his question—how does a bony fish adjust the salt content of its cells so that it can move from one part of the ocean that is salty to another that is less salty? Is osmosis (the flow of water from an area of low salt concentration to one of higher salt concentration) an active or a passive process? Does something happen in the fish to make osmosis active? He published his findings in "Variation de la pression osmologique du sang des poisons téléostéens" (Variation in osmotic blood pressure in teleost fish), in *Comptes rendus des séances de la Société de Biologie et des ses filiales* (1922). He varied the fish used and continued this experiment for several other species.

Acclaim by his university came in the form of a laboratory. In 1923, a chair in comparative physiology was established for him at the University of Paris. He mounted a major research effort described in *Bulletin de l'Academie de Médecin*, on the essential role of carbon dioxide in cell regeneration and synthesis. Portier's next project was one with ecological implications. He researched the cause of death among birds befouled by oil spills, concluding that they die of heat loss. This 1934 work has made a major difference in how such industrial accidents are dealt with, particularly in areas of endangered wildlife.

Ever energetic and bright, Portier continued working into his 80s, producing his last work, *The Biology of Butterflies*, in 1949. He continued to be active in both the Academy of Medicine and the Academy of Sciences, serving on committees of both bodies. He died at home in Bourg-la-Reine on January 26, 1962.

⊠   Pourtalès, Louis-François de
(1823–1880)
American
Biologist

There is some doubt about the year of Pourtalès's birth. His birthdate is reported to be either March 4, 1823, or 1824; and his birthplace to be Neuchâtel, Switzerland. He was a young student who went to study with LOUIS AGASSIZ in Switzerland in 1847 when Agassiz settled in Cambridge, Mas-

sachusetts. He does not seem to have attended a university, but instead served as Agassiz's apprentice. The collaborative work written by Agassiz and Gould, *Principles of Zoology*, published in 1848, was illustrated by Pourtalès. In the same year, Pourtalès joined the U.S. Coast Survey. His talent for accurate drawing led to his rise to head of the tidal division by 1854 and a major role in coastal surveying.

Although surveying coastal waters was his occupation, research was an integral part of it. A directive from ALEXANDER BACHE, prominent physicist who was head of the Coastal Survey, required all specimens brought up when taking soundings to be preserved. Soundings were taken very primitively. A weighted line was dropped and measured when it hit ocean bottom. That point is usually obvious since the line goes slack. Inaccuracies were frequent because the line has to be heavily weighted or it will not sink to great depths. If it became entangled, the reading was false.

Most of the specimens were sent to Jacob W. Bailey in West Point, New York, for analysis, but Pourtalès was the expert on foraminifera, unicellular aquatic organisms that create very complex shells (tests). The tests are composed of either calcium compounds or chitin, the polymer that is found in shrimp, lobster, and crab shells. He had data for over 9,000 samples showing bottom sediments from Massachusetts to Florida. That collection dated from work done with Louis Agassiz who had, at Bache's suggestion, collected coral in Massachusetts waters. This had led Agassiz to travel south in 1851 with Pourtalès as his assistant on a collecting trip sponsored by the U.S. Coast Survey. They explored the coral reefs near the Florida Keys, and Pourtalès collected an array of sepunculids and holothurians. Sepunculids, or peanut worms, are a phylum of marine organisms that may be related to segmented worms. Holothurians, or sea cucumbers, are armless relatives of starfish.

After the Civil War (1861–65), Benjamin Pierce, a friend of Agassiz, provided transport for Pourtalès on the *Corwin* in 1867 and the *Bibb* in

1868 and again in 1869. His objective was to study deep water and its animal content. To do this, Pourtalès developed technology for dredging deeper waters than had been done before. He designed equipment that functioned at depths of 850 fathoms (820 m or 2,700 ft), which he used off the east coast of Florida. This improved collection and led to greatly increased knowledge of the geology and zoology of this undersea region. This significant work was published in the *Report of the Superintendent of the U.S. Coast Survey for 1869, Appendix 10*. Agassiz is listed as the sole author. Pourtalès published his book on the subject, *Deep Sea Corals*, in 1871. His other special interests were in the crinoids—the sea "lilies" that are really animals—and the holothurians, or sea cucumbers.

Another more ambitious exploratory cruise in the North Atlantic was planned for 1871–72 aboard the Coast Survey's ship, the *Hassler*. Agassiz was in charge of the whole expedition with Pourtalès in charge of dredging. The voyage was not a success; the equipment did not function as planned. However, it did inspire the British, notably WILLIAM CARPENTER, a marine scientist, and others, to redouble their efforts in promoting the HMS *Challenger* voyage. The *Challenger* expedition was a major undertaking. The ship sailed in 1872 with a complement of scientists who would spend the next three and a half years studying the world's oceans, the organisms found in them, the muds and oozes that were collected from the bottom, and the weather encountered. Thus, the success of the *Challenger* was a direct outgrowth of the *Hassler* failure.

After the death of his father in 1870, Pourtalès settled his affairs in Switzerland and was financially independent. He then left the Survey in 1873, settling near Cambridge, Massachusetts. Louis Agassiz vacillated for years; he could not make up his mind about his successor as curator of the Museum of Comparative Zoology (the Agassiz Museum): his son, Alexander, or his protégé, Pourtalès. After his death in December 1873, the two settled it: Alexander Agassiz would be curator and Pourtalès the keeper, or administrative head. He maintained that position until his death at home in Beverley, now part of Salem, Massachusetts, on July 17, 1880.

Unfortunately, much of Pourtalès's work remained unfinished. He did not have the time to write it all up, and some of his earlier work was not credited to him. He was, however, a major figure in the development of scientific interest in the coastal waters of the United States and an active member of both the National Academy of Sciences and the American Academy of Arts and Sciences.

Q

⊠ **Quatrefages de Bréau, Jean-Louis-Armand de**
(1810–1892)
French
Biologist

This much-traveled scientist was born in Valler-augues, France, on February 10, 1810, to Jean-François and Marguerite-Henriette (de Cabanes) Quatrefages de Bréau. In spite of his religion (he was a Protestant), the young Quatrefages was accepted as a student in the Collège Royale in Tournon where he studied mathematics and science from 1822 to 1826. Leaving this secondary school, he then attended lectures in the University of Strasbourg where he defended the two theses required for the doctorate in sciences, one on lunar volcanic activity in 1829, the other on exploding cannons in 1830. The university gave him an appointment as the assistant for chemistry and physics. While serving in that role, he wrote yet another thesis for the faculty of medicine, in 1832, on the extrusion of the bladder. Based on his work in the medical area, he practiced medicine in Toulouse beginning in 1833, and while there, founded the *Journal de médecine et de chirugie de Toulouse* (Journal of medicine and surgery of Toulouse) in 1836. His displeasure at his appointment to the faculty of medicine in the zoology department prompted his departure for Paris in 1840.

French naturalist and anthropologist Jean-Louis-Armand de Quatrefages de Bréau. *(Hulton/Archive/Getty Images)*

In Paris, Quatrefages pursued another doctoral degree, this time in natural sciences. While doing so, he supported himself by writing for the popular press and drawing the illustrations for GEORGES CUVIER's work *Le regne animal* (The animal kingdom), which HENRI MILNE-EDWARDS was preparing for publication. Quatrefages concentrated on the invertebrate entries, and for 15 years, starting in 1840, wrote more than 80 articles and made a number of trips to the Atlantic coast to collect specimens. In 1844, he accompanied Milne-Edwards on a collecting trip to the Sicilian coast. The diary of this exploration was published as *"Souvenirs d'un naturaliste"* (Memoirs of a naturalist) in 1854. It was a popular success. That same year, Quatrefages was appointed to the chair of natural history at the Lycée Henri IV. This was and still is a prestigious secondary school. He was appointed to the anatomy and zoology sections of the Academy of Sciences in 1855.

The national disaster of silkworm disease, which devastated the textile industry in Lyon, was the arena in which Quatrefages came to national prominence. He worked on this problem from 1857 to 1859. By 1865, on Milne-Edwards's recommendation, Quatrefages became chairman of the department of the natural history of man in the Muséum d'histoire. This ended his career in natural science, but he did produce a significant work on the natural history of annelid worms in marine and fresh waters in 1865.

Scientifically, Quatrefages followed the lead of Milne-Edwards in the study of "organic complication" or "degradation" of organisms. His study of marine worms gave him much material to work on. Using this data, he generated theories explaining the appearance or disappearance of organs, and the transfer of their function to other structures. These ideas were of some interest and created considerable discussion, but have since disappeared from the body of scientific knowledge. This idea is now only a historical artifact. We know that it created considerable interest, and many writers commented on it in its day. However, organs do not "degrade." Different organs in different animals can perform analogous functions without being related.

Although Quatrefages was the first to study comparative histology, he never accepted cell theory as applied to animals. Neither did he accept the ideas of CHARLES DARWIN, the biologist who introduced the concept of the evolution of all organisms from a single source. Quatrefages maintained a belief in a totally separate creation for man. He was a scientist whose name was recognized by the average thoughtful reader, and was a member of many French and foreign scientific societies as well as a founding member of the French Association for the Advancement of Science. He died in his Paris home on January 12, 1892. By then, much of his theoretical writings had been discarded.

R

⊠ Rathbun, Mary Jane
(1860–1943)
American
Invertebrate Zoologist

The youngest of the five children of Charles Howland and Jane (Furey) Rathbun was born in Buffalo, New York, on June 11, 1860. She was brought up by her aunts after her mother died when she was a baby. Charles Rathbun was a stonemason and all his children grew up wandering through his quarry and picking fossils out of the rock. Rathbun's formal education ended when she graduated from Buffalo's Central High School in 1878. Her career opportunity came as a result of a summer vacation.

Mary Rathbun traveled to Woods Hole, Massachusetts, in the summer of 1881 to spend some time with her brother Richard, who was then an assistant at the U.S. Commission of Fish and Fisheries (known familiarly as the Fish Commission). She volunteered to clean, sort, and classify specimens, and proved to be very good at the task. By 1884, Spencer Baird, Commissioner of the Fish Commission and Secretary of the Smithsonian Institution in Washington, D.C., appointed her to a full-time clerk's position. At the time of Rathbun's work for the Fish Commission, a clerk's position was one of some standing. It meant being promoted from the ranks of sorters.

Many other women, called "Baird's girls," were sorters. Because it was difficult for women to find employment, this was a desirable, albeit dull, job.

Baird was a natural connection between the Fish Commission and the Natural History Museum, a part of the Smithsonian in Washington. The museum created a Department of Marine Invertebrates in 1881, with Richard Rathbun as its head, and in 1886, Mary Jane Rathbun was appointed as his assistant. Richard Rathbun traveled considerably, visiting a variety of collection sites. His sister was the coordinator of the department's sorting, classifying, and labeling of specimens, and its official correspondent. She also found time for research; her area of specialty was the Decapoda (shrimp, lobsters, crabs, and the like), and she used both specimens she had collected at Woods Hole and in Maine, and others that she obtained by exchanging with other collectors and museums.

Her first paper, "Genus Panopeus," was coauthored with J. E. Benedict, and published in *U.S. National Museum Proceedings . . .* in 1891. By 1894, Rathbun was an aide in the department, and in 1898, she was the second assistant curator. As such, she was one of four professional women at the Smithsonian Institution. In 1896, Rathbun traveled to Europe, visiting museums that had significant invertebrate collections. The resulting publication was a monograph, *Les crabes d'eau, douce, Potamonidae* (Freshwater crabs, Potamonidae). This work appeared in *Nouvelle archives, Paris, Musèe d'histoire naturelle* (New archives, Paris, Museum of Natural History) Nos. 6, 7, 8 in 1904, 1905, and 1906.

In 1907, Rathbun was appointed assistant curator in charge of the function of the Division of Marine Invertebrates (as her department was renamed); seven years later she arranged her resignation—Rathbun had come into an inheritance, so she no longer needed the financial support of the position, and she wanted more time for research. Her intent was that her assistant, Waldo Schmitt, succeed her. He did. This move gave her the position of honorary curator with access to the research facilities and no administrative duties to attend to. The end result was the completion of her definitive work on the crabs of North America, all published in the *Bulletin of the U.S. National Museum:* "Grapsoid Crabs of America" (1918), "Spider Crabs of America" (1925), "Cancroid Crabs of America" (1930), and "Oxystomatous Crabs of America" (1937), and a work on extinct animals, "Fossil Stalk-Eyed Crabs of the Pacific Slope of North America," published in *U.S. Geological Survey* (1935). During this period of extensive writing, Rathbun also acted as a consultant for petroleum geologists who were amassing indicator fossils. An indicator fossil is one that is characteristic of a particular rock formation and/or of a specific geologic period. Finding such a specimen dates the rock that surrounds it and often specifies where that rock came from.

Rathbun was an active researcher for almost all of her adult life, with one period of absence. From 1916 to 1918, she worked for the Red Cross. In her research time, she identified 1,147 new species and 63 new genera, and was instrumental in stabilizing the nomenclature in this exploding field. She was awarded an honorary M.A. by the University of Pittsburgh in 1916, and a Ph.D. in biology by George Washington University in 1917. Her memberships included the American Association for the Advancement

of Science, and the Washington, D.C., Academy of Science. Rathbun retired again in 1938; she died in Washington, D.C., on April 4, 1943, one of the two of "Baird's girls" who spent an entire working life in research.

⊠ **Renard, Alphonse**
(1842–1903)
Belgian
Geologist

Alphonse Renard was born in Renaix (now Ronse), Belgium, on September 26, 1842. While still a child, he was sent to the seminary; his family had chosen him for the priesthood. His first introduction to science came at Maria Laach, a Jesuit school in the Eifel, a geologically fascinating plateau north of the Mosel River and east of the Ardennes, which is part of the Rhenish Slate Mountains. The region is characterized by deep and steep valleys, crater lakes, and extinct volcanoes. Renard spent his leisure time climbing and collecting rock specimens. In 1874, he became professor of chemistry and geology at the Jesuit *collège* (secondary school) in Louvain. Three years later, he was ordained a priest.

Renard's next position was a secular one; he was appointed as a curator of the Royal Natural History Museum in Brussels, and by 1882, had totally given up teaching duties. At this point in his life, Renard was experiencing considerable personal difficulties. His increasing involvement with scientific research was affecting his position in the Jesuit order, and he finally withdrew from the order in 1884. In 1888, the University of Ghent offered him the chair of geology, and he remained in this post for the rest of his life.

Renard's reputation as Belgium's foremost geologist began in 1874 with his study of the plutonic rocks of the Ardennes. It was obvious to him that these formations were of deep-sea origin—plutonic rocks are those that originate on the ocean bottom. This work was first published in the *Memoirs of the Royal Academy of Arts and Sciences of Belgium*, and then as a book with Charles de La Vallée-Poussin as coauthor.

Renard continued to write on geological events that produced deep-sea rocks, notably the metamorphosis of igneous rocks, those formed by volcanic eruption. His reputation as a major European figure in geology made him an ideal candidate to examine the specimens brought back by the *Challenger* voyage, the globe-encircling, three-and-a-half-year-long expedition (1872–76) of the HMS *Challenger*, the ship commissioned by the British Navy to study the ocean in all its aspects.

Working closely with JOHN MURRAY, Renard published several preliminary works, and then, with Murray as coauthor, wrote the book *Deep-Sea Deposits*. This magnificent work was published in 1891. It opened up vast new areas of oceanography and was a part of a series of monograph reports of various aspects of the *Challenger* voyage. Prior to the evidence of the *Challenger* specimens, the ocean bottom was assumed to be a lifeless, featureless wasteland. The book detailed the geographic distribution of the various sediment types found in the deep ocean: red clays; globigerina, radiolaria, diatom, and pteropod oozes; dust, both terrestrial and cosmic; rock and mineral concentrations. The location and composition of the manganese nodules the expedition found are all described in detail.

Some of Renard's scientific papers specifically dealing with marine science are "On Nomenclature, Origin, and Distribution of Deep Sea Deposits," published in *Proceedings of the Royal Academy at Edinburgh* (1884), which was also published in French in *Bulletin of the Royal Natural History Museum of Belgium;* "On Microscopic Characters of Volcanic Ashes and Cosmic Dust . . ." published in *Proceedings of the Royal Academy at Edinburgh* (1884), with John Murray as coauthor; and "Preliminary Notice on the Marine Sediments Collected by the *Belgica* expedition. . . ." published in *Memoirs of the Royal*

Academy of Arts and Sciences of Belgium in 1901 and 1902.

Renard, who married in 1901, wrote a number of articles in which he tried to bridge the gap between formal religion and science. His breach with the Jesuits left him unhappy but he continued researching and writing on science until his death in Brussels on July 9, 1903.

The work done by Renard is significant because even before the evidence brought back in the *Challenger* specimens, he was pointing out the deep-sea origins of old terrestrial rock formations far from the sea and producing cogent theories to explain their origin.

Revelle, Roger Randall Dougan
(1909–1991)
American
Geophysicist, Meteorologist

Roger Revelle was born in Seattle, Washington, on March 7, 1909, to William and Ella (Dougan) Revelle, both graduates of the University of Washington. The family moved to Pasadena, California, in 1917, and Roger Revelle attended local schools before entering Pomona College. Originally, he was enrolled as a journalism student, but a lecture by Alfred Woodford caused him to change his mind and major in geology. While an undergraduate, he met Ellen Clark, an undergraduate at the newly opened Scripps College and a grandniece of the founder of her college. Revelle graduated from Pomona in 1929, stayed on for a year of graduate study, and later married Ellen Clark in 1931.

By 1930, Revelle was in the graduate program at the University of California, working with George Louderback, whose area of interest was sedimentation. That year, the Carnegie Institution in Washington, D.C., requested that several core samples be examined at the Scripps Institution of Oceanography, and prompted the director, Thomas Vaughan, to ask Louderback for

Roger Revelle with Nansen bottle on Scripps Institution of Oceanography Pier. *(SIO Archives/UCSD)*

a promising graduate student. He recommended Revelle, who accepted the position. In 1931, the Scripps Institution was a small operation (one laboratory, one research ship, and a marine station) that was attached to the University of California. Oceanography was then very much a European science.

Working on the Pacific sediments collected by the *Carnegie*, a research vessel operated by the Carnegie Institution, Revelle and his coworkers found that only half of the carbon dioxide released by fossil fuel went into the sea, not almost all, as had been previously believed. This painstaking effort began in 1930 and continued

for some years. Revelle collected marine samples for both Carnegie and Scripps, comparing the samples from different oceans. It was the beginning of his continuing interest in the carbon cycle and the measurement of atmospheric carbon dioxide. He made it the subject of his dissertation, "Marine Bottom Samples Collected in the Pacific Ocean by the *Carnegie* on its Seventh Cruise." The University of California accepted it in 1936 and Revelle was granted a Ph.D. In the same year, HARALD SVERDRUP came to Scripps as director, and he moved its research direction definitively into modern physical oceanography. The Revelle family moved to Bergen in Norway, in 1936, for a year of study with BJØRN HELLAND-HANSEN at the Geophysical Institute.

En route to Norway, Revelle stopped in Washington to meet with members of the U.S. Navy Hydrographic Office to discuss the *Bushnell* Report. Earlier, in 1935, Revelle had been on the submarine USS *Bushnell* as it cruised the Pacific, and he tried to interest the navy in supporting scientific voyages. Revelle's first stop in Europe was the meeting of the International Union of Geodesy and Geophysics in Edinburgh, Scotland. This was his introduction to the world's leaders in the field. The year in Bergen gave Revelle additional sea time on the institute's vessel, the M/S *Armauer Hansen*.

Returning to California, Ellen Revelle came into her inheritance of the newspaper fortune amassed by her grandfather, James Scripps, and Roger Revelle taught marine geology at Scripps and physical oceanography at the University of California at Los Angeles. He was also given a grant by the Geological Society of America to study Pacific Ocean sediments. While working on that, he explored the Bay of California on the research vessel *E. W. Scripps*. At the same time, Revelle met WALTER MUNK, then a graduate student, who became a lifelong friend and coworker.

Early in 1941, Revelle, who was in the Naval Reserve, was called for sonar training. Sonar is an acoustical technique involving bouncing sound waves from a ship down to the ocean bottom. The returning echo is informative; its return time indicates the depth of the ocean or the presence of something between the ocean bottom and the sonar-equipped ship. The layering of water is dependent on what is in it. The return of the sonar signal can be delayed by a number of things, and one is a layer of "dense water" that diffuses but does not stop the sonar echo. Revelle was involved in research on radar propagation and the use of sonar in the U.S. Navy's San Diego base. In Washington, at the navy's Hydrographic Office, he researched the deep, scattering layer and sonar analysis of it.

His work with the navy was enlarged to include research in many areas other than sound; it grew to include the study of ocean waves and surf, the effects of these on landing craft, the effects of winds and currents on small boats or rafts, and the effects of water density differences on submarines. All of these were directly involved in the navy's war effort in World War II. By 1945 Revelle, who had risen to the rank of commander, was part of the staff of amphibious forces of the Pacific Fleet and involved in the planning for the invasion of Japan.

When the war ended, Revelle was assigned to Operation Crossroads. This was the joint military command to assess the effects of the atomic bomb test on Bikini Atoll in 1946. Revelle was head of the oceanographic and geophysical divisions of that group. The second survey of Bikini in 1947 took drill cores on the island to a depth of 800 m (2,640 ft) to find sediments that were about 30 million years old. The sediments were reef limestone laid down in shallow water. They proved that CHARLES DARWIN's theory of reef-building was correct. Atolls are volcanic islands that sank and were then overlaid with many generations of coral skeletons.

Revelle was transferred to the Office of Naval Research while still part of Operation Crossroads and, working with Sverdrup, he persuaded the navy to expand its research efforts.

The result was the organization of the Marine Physical Laboratory, eventually part of the Scripps Institution, and the acquisition of two refitted naval vessels as research ships. The State of California also contributed funds; the combined effort moved this small research organization to the ranks of major oceanographic research stations. Sverdrup wanted to go back to Norway, and he proposed Revelle as his successor at Scripps. After a prolonged search and much argument, Revelle became the associate director and professor of oceanography in 1948 and director in 1950.

The Scripps Institution greatly expanded its work in the 1950s. It was obvious that a graduate center was needed in La Jolla, California, and Revelle worked to establish that. A good deal of research was being done at La Jolla, and to regularize it and train a new group of researchers, a graduate faculty and a university structure were necessary. The University of California, San Diego, is a direct result of his efforts. Revelle was very involved in the research of Scripps, organizing and sailing on the *Northern Holiday* (1951), *Shellback* (1952), *Capricorn* (1952), *Transpac* (1953), *Norpac* (1955), *Downwind* (1957), and *Naga* (1959). These voyages produced a number of important finds: just how thin the layer of bottom sediments was, the considerable flow of heat up from the sea bottom, and, moving through the ocean's water, the very recent (geologically speaking) appearance of seamounts, the existence of a mid-ocean ridge in the Pacific, and the massive cracks in the ocean bottom perpendicular to the mid-ocean ridges—the transform faults. This data did not fit with theories then current of geologic history. Revelle therefore urged others to expand geophysical research to the sea, to examine the phenomena of heat flow and geomagnetism. The resulting data became part of the evidence that led to the concepts of seafloor spreading and plate tectonic theory.

In 1957, Revelle and Hans Suess published an article in *Telus* on atmospheric carbon dioxide,

"Carbon Dioxide Exchange Between Atmosphere and Ocean and the Question of an Increase of Atmospheric Carbon Dioxide During the Past Decades." This problem, the increase in atmospheric carbon dioxide, was one that he continued to work on for the rest of his life. When he was awarded the National Medal of Science in 1990, he told a reporter that he got the award for being the "grandfather of the greenhouse effect."

Although Revelle gained an international scientific reputation for his oceanographic work, he became better known in the general scientific community and among the public as a spokesman for science in his position at the National Academy of Sciences. He had been elected a member in 1957. He was also a member of the U.S. National Commission for the United Nations Educational, Scientific and Cultural Organization (UNESCO) and helped plan its Office of Oceanography. He devoted considerable time and effort to the U.S. National Committee on the International Geophysical Year (1957–58), to the study of the environmental effect of radiation, and to scientific adviseries to Congress. Revelle was instrumental in the launching of the first International Oceanographic Congress in New York City in 1959. This meeting sparked the discussion that eventually culminated in the International Indian Ocean Expedition, a major undertaking that aimed to study the Indian Ocean in all its aspects. Several studies were launched in the early 1960s, and there have been several international exploratory voyages since then, notably in the 1990s.

Revelle spent much of the early 1960s in Washington, D.C., as adviser to the Secretary of the Interior and a member of the Federal Council for Science and Technology. Accepting a professorship at Harvard in 1964, he became the Saltonstall Professor of Population Policy and the first director of the Center for Population Studies. The center's object then was the study of the interaction of growing population, resources, and the en-

vironment. He continued to travel extensively, actively involved in oceanic and atmospheric science, international cooperative scientific ventures, and science policy. He was president of the American Association for the Advancement of Science in 1973, and worked on various committees of that organization. Revelle returned to the University of California, San Diego, in 1976 as professor of science and public policy. He divided his time between there and Harvard for two years and then left Harvard. From 1985 to 1988, Revelle also served on the National Aeronautics and Space Administration Advisory Council.

In addition to the National Medal of Science, Revelle was awarded the Tyler Ecology Energy Prize in 1984, and the 1986 Balzan Foundation Prize, which was given in recognition of his life in science. Roger Revelle died on July 15, 1991, and four years later the Scripps Institution launched a new research vessel named the *Roger Revelle*.

Revelle's life in science touched many fields, all connected to his love of the sea. He was an excellent researcher, teacher, mentor, academic administrator, and public servant. He took the last title seriously and devoted much energy to making the world a better place for people in the present and preserving it for the future.

⊠   **Richet, Charles Robert**
(1850–1935)
French
Physiologist

Charles Richet was born in Paris on August 26, 1850, son of Alfred Richet, a distinguished surgeon who worked in Pierre Berthelot's laboratory. Berthelot was a famous French chemist whose area of expertise was the synthesis of organic compounds such as alcohols. He was the discoverer of the methodology of the uptake of nitrogen by plants and, pursuing that, he established the steps in plant synthesis of protein. This is plant physiology.

Charles Richet *(Hulton/Archive/Getty Images)*

Trained in medicine, Charles Richet went on to specialize in physiology at a time when that science was almost exclusively a German province. By 1887, he was chair of physiology at the faculty of medicine in Paris. Richet did significant work on microorganisms and the use of toxins to create disease resistance. One of the diseases that he spent much time studying was tuberculosis. He did not find a cure, but he was truly expert in the life cycle of the bacillus. Richet was also well aware of the toxicity of the blood of one animal to that of another.

A renowned researcher, he was invited, along with PAUL PORTIER, by ALBERT I to investigate the toxicity of *Physalia*, the Portuguese man-of-

war, and some sea anemones. They were quite successful, isolating the toxic material, analyzing it, and determining lethal dosages by varying the subject animals. These results were presented in 1902 at the meeting of the Société de Biologie. This began a serious study by Richet and his coworkers of anaphylactic reactions (shock); the researchers found that other animals such as some mussels species can also induce this violent autoimmune response. This discovery redirected physiology, for it became necessary to explain why a harmful reaction that can kill the organism, anaphylaxis, is helpful to the species. It is helpful because it preserves the chemical "identity" of the species. This was as close to an understanding of the unique quality of species DNA as Richet could achieve without knowing about DNA. He was rewarded for his singular work with the 1913 Nobel Prize for physiology and medicine.

Richet wrote numerous papers that were published for the most part by *Comptes rendus. . . . de la Société de Biologie* and *Annales de l'Institut Pasteur*. He also wrote plays, poetry, general history and, in a society that was determined to go to war, pamphlets supporting the pacifist cause. During World War I (1914–18), Richet worked on the problems of maintaining and delivering blood plasma and transfusions. For this effort, in 1926 he was awarded the Croix de Guerre by the Legion of Honor. Armed with more information about blood, Richet continued his work on transfusion. He died in Paris on December 3, 1935, burdened with the knowledge that war was coming again.

Rondelet, Guillaume
(1507–1566)
French
Biologist

Guillaume Rondelet was born in Montpellier, France, on September 27, 1507, a younger son in the large family of a drug and spice merchant. He was orphaned as a young child and reared by an older brother. In 1525, he left the south of France for Paris, where he was a student in the Faculty of Arts and Letters (Sorbonne). He left Paris with that degree in 1529 and attended the Medical Faculty at Montpellier. This is an ancient school and claims to be the oldest in Europe. By the next year, he was procurator—a post similar to that of a laboratory assistant—and a known figure in the university community. It was then that he met and was befriended by François Rabelais, the physician, raconteur, and author of *La vie de Gargantua et Pantagruel*. (Rabelais's fictional character, Rondibilis, is essentially a portrait of Rondelet.)

To support himself, Rondelet taught in secondary schools and practiced medicine in Vaucluse until, in the mid 1530s, he returned to Paris to study anatomy. He finished his doctorate in medicine in 1537 and returned once more to Montpellier. The next year, he married Jeanne Sandre. They were supported by her family until 1545 when Rondelet was appointed Regius Professor of Medicine.

Beginning in 1540, and overlapping with his appointment at the Medical Faculty, Rondelet was the personal physician of Cardinal François Tournon and traveled widely with him in France and what is now Belgium. On one of these long trips, in 1549, the cardinal's retinue visited Rome and then several universities in northern Italy where Rondelet met many scholars that he had known through their correspondence.

When Rondelet left the cardinal's service in 1551, he returned to Montpellier, to his collections of plants and animals, and his lectures. He was elected chancellor of the university in Montpellier in 1556 and while in that position, he was the prime mover for curriculum reform and the institution of the anatomy theater there.

Ever popular as a lecturer in anatomy and natural science, Rondelet drew students from all of Europe; one of his most famous students was Conrad Gesner, the Swiss naturalist whose encyclopedic work on plants and animals known to

Europe included the first illustrations of fossils. Rondelet's great work was written between 1554 and 1556. It is *Libri de piscibus marinis in quibus verae piscium effigus expressae sunt. Pt. 2. Universae aquatilium historial pars altera*. This was translated into French by another student, Laurent Joubert, as *L'histoire entière des poissons* (The life history of all fish) in 1558. This massive, illustrated work dealt with all aquatic animals, both freshwater and marine. Among other things, Rondelet realized that dolphins and otters were mammals, not fish, and that those animals that lived in rivers could not make the transition to the sea and live. This work became the standard reference and zoological text dealing with aquatic organisms and remained so for over a century.

Late in his life, Rondelet seems to have either become a Protestant or was interested in their thinking. He retired to Realmont (Tarn) months before he died there on August 30, 1566.

⊠   **Ross, Sir James Clark**
(1800–1862)
British
Physicist

Born in London on April 15, 1800, James Clark Ross, third son of George Ross, entered the Royal Navy as a 12-year-old midshipman. He was assigned to his uncle's command; John Ross was by then a well-known navigator and explorer. The object of the Ross expedition was surveying the North Sea. John Ross's assignment continued when the expedition went on to survey and chart the White Sea in 1815–17. One notable feat on the latter voyage was the determination of the longitude of Archangel—the northern Russian port—using a transit of the planet Jupiter by one of its moons as the fixed astronomical point.

Months after their return to England in 1818, John and James Ross, together with William Parry as second-in-command, sailed to North America in an attempt to find the fabled

Northwest Passage by sailing up the Davis Strait between Greenland and Baffin Island. This was yet another failure to find the Northwest Passage, but on their return, the same three men were chosen for another expedition to the Canadian Arctic. This plan to find the Northwest Passage was to force their way through Lancaster Sound in the Northwest Territories and the Bering Strait. This voyage in 1819–20 met with some success and the team was awarded a £5,000 prize from the Board of Longitude. This then financed the next attempt and insured the promotion of James Clark Ross to lieutenant in December 1822. Before and after his promotion, James Ross and William Parry had been charting Arctic waters in the HMS *Fury*. They were on that ship on another Arctic charting voyage when they were wrecked in July 1825 and rescued by the HMS *Hecla*.

This experienced team then attempted in 1827 to sail to Spitsbergen, off the northern coast of Norway, and cross to the North Pole on the ice. This too did not work, but Ross was promoted to commander in November of 1827. Two years later, both John and James Clark Ross embarked on another expedition in the Canadian Arctic. This trip was financed by Felix Booth (the gin manufacturer), and so they named a large, newly charted part of the Arctic Boothia Peninsula in his honor, and a large island for the king, William IV. While on this voyage, James Clark Ross took frequent compass readings every day and established the location of the North Magnetic Pole. It was then at 70°.05'17" N and 96°45'48" W. The voyage ended in 1833 and Ross was promoted to post captain the next year.

Ross sailed again in Arctic waters when he was on the HMS *Cove* in 1836, rescuing icebound whalers. His next trip—on land—was a purely scientific one. In 1838, he was selected to do a magnetic survey of the United Kingdom for the British Association for the Advancement of Science. The next expedition for this much-traveled man was directed to the Antarctic. It lasted from 1839 to 1843. While charting Antarctic waters, Ross discovered the only active volcano on that continent on New Year's Day, 1841. He named it Mount Erebus. He also charted and named Victoria Land. The scientific studies he engaged in on this voyage involved the determination of the location of the South Magnetic Pole, geology, and sampling of marine life at varying depths. The voyage was a grand success; Ross returned with a wealth of sea life specimens and had lost only one man to illness. This voyage ended the seagoing segment of his career.

After his return from Antarctica, Ross continued in the Royal Navy and concentrated on writing his account of the last voyage. It was a two-volume work published in 1847, entitled *A Voyage of Discovery and Research in the Southern and Antarctic Regions During the Years 1839–1843*. His collected work on magnetism was published by Edward Sabine in *Philosophical Transactions of the Royal Society of London* in 1854.

Ross was an early and avid collector. When on his first cruise with his uncle, he had gathered botanical specimens and was elected a member of the Linnaean Society in 1824. By 1828, he was a fellow of the Royal Society. On his return from the Antarctic voyage, he received medals in 1842 from the Geographic Societies of London and of Paris. In 1844, he was knighted by Queen Victoria and received the French Legion of Honor rosette. He maintained memberships in the Paris Academy of Science and other foreign scientific societies. James Clark Ross died in Aylesbury, England, on April 3, 1862.

S

Sabatier, Armand
(1834–1910)
French
Comparative Anatomist, Zoologist

Armand Sabatier was born near Montpellier on January 14, 1834. His family was one of devout Protestants. Educated locally as a child, he attended the University of Montpellier, then enrolled in the Faculty of Medicine. This prestigious institution is the oldest medical school in Europe. Upon graduation in 1855, he was appointed as an assistant in anatomy. Three years later, he served an internship in Lyons but returned to Montpellier to finish his doctoral thesis. Since he was not a full-time student and had departmental teaching duties, this went on for some time. He was appointed assistant professor in 1869. For the Franco-Prussian War of 1870, every available medically trained person was summoned, and Sabatier served with the medical personnel during this conflict.

Upon his return to Montpellier, he continued teaching. In 1873, he was promoted to associate professor, and in 1876, to professor of zoology in the Faculty of Sciences. His dissertation, *Etudes sur le coeur et la circulation central dans la série des vertèbres* (Studies on the heart and the central circulation in the series of vertebrates),

was published simultaneously in Paris and Montpellier in 1873. Sabatier's specialty was cardiac morphology, the physiology of amphibians and reptiles, and the evolution of the heart. The last is significant since the heart differs in the series of vertebrates from fish to amphibians, to reptiles, to birds, and to mammals.

Best known for his work on the embryology of invertebrates, Sabatier published extensively on various aspects of the physiology of mussels and on reproduction in invertebrates. In his time, this was a totally new field. Sensing the need for a marine laboratory, he worked for years to amass sufficient funds for such an installation. The Station Zoologique de Sête was finally begun in 1879, and its maintenance required constant fund-raising. While doing that, Sabatier also trained many students, teaching the techniques in microscopy that he had developed.

Sabatier's publications include "Sur quelques points de l'anatomie de la moule commune" (On several points in the anatomy of the common mussel), "Études sur la moule" (Studies on the mussel), "Formation du blastoderme . . ." (Blastoderm formation), "La spermatogenèse chez les annelids . . ." (Spermatogenesis in annelids) and "De l'ovogenèse chez les ascidiens" (Ovogenesis in ascidians). He seems to have worked on several lines of research simultaneously and published all of the works cited in the 1890s shortly before his retirement from formal teaching. They were all very long articles in either the *Comptes rendus . . .* , the publication of the National Academy of Sciences, or the *Memoires de l'Academie . . .* the publication of the Academy of Sciences at Montpellier.

These works and others on similar subjects gained for Sabatier an international reputation as the expert in the field. Until he began publishing this work, the reproductive cycles of many marine organisms were totally unknown.

Sabatier also published a number of works, between 1900 and 1910, in which he tried to explain his view of both science and faith. Although an evolutionary biologist and a devout Christian, he did not find conflict in either belief. His publication "Philosophie de l'effort" (Philosophy of effort) expounds his belief that survival of a species depended on the striving for a moral life. Surrounded by students who loved him, he continued his work and teaching until his death in Montpellier on December 22, 1910.

⊠ **Sars, Michael**
(1805–1869)
Norwegian
Marine Biologist

Sars was born on August 30, 1805, in Bergen, Norway, son of Michael Sars, a sea captain from Bremen, Germany, and Divert Heilman, originally from Estonia. The family moved to Christiania, now Oslo, and the future biologist started his academic career in theology. He became a pastor in 1828 and went to a small fishing and farming village in western Norway; zoology was a hobby. Sars married Maren Welhaven in 1831 and they had 12 children; one son, Georg, became a biologist. Sars wrote articles on his marine finds and was favorably noticed by the scientific community. This resulted in a government grant that financed a six-month trip to Holland, France, Germany, Denmark, and Sweden, to observe the methods and practices of other marine zoologists.

The area of study that occupied Sars for 30 years, 1830 to 1860, was the life cycle of marine invertebrates. Many juvenile forms of marine animals look very different from the adults. On numerous occasions, they had been classified as entirely separate genera. Sars made several great contributions to early marine biology. One was the result, in 1830, of using Otto Muller's dredge. With it Sars came up with many sensational finds from the deep ocean bottom. His finds defied the generally held assumption of this period that the deeper one went in the sea, the fewer species and the fewer individuals of those species one would find. Sars was one of the first to show the scientific world that this assumption was untrue. Another example of his innovation was the careful work revealing that great change in form occurred on maturation, and that in some, notably the coelenterates, there was alternation of generations. Hydroid forms that are sessile (they do not move) alternate with free-swimming medusal (jellyfish) forms.

After the first trip in 1837, Sars put theology aside and worked as a biologist for the rest of his life. Working with SVEN LOVÉN, in 1837 he studied the larval stages of annelids. Annelids are segmented worms; the most familiar terrestrial annelid is the earthworm. Three years later, the two researchers used similar techniques to study the life cycle of mollusks. In these animals, the larval stages do not look like the adult animals at all. The larval and adult forms of starfish and other echinoderms were the next subjects of this analysis in 1844 and 1846. By then, Sars had an international reputation. His renown led to a government-sponsored exploratory voyage to the Adriatic Sea in 1851 and another in 1852–53 to southern Italian and Sicilian waters. Part of this latter voyage included a number of stops to meet other European zoologists.

The University of Christiania (Oslo) declared Michael Sars professor of zoology *extraordinarius* in 1854. While a great researcher, he was an impatient teacher and not a popular one. By 1864, Michael Sars was sailing and working with his son Georg Sars. That was the year when Georg found living crinoids. These echinoderms were thought to have been extinct since the Mesozoic era. The two continued deepwater work; in 1868 they collected 427 specimens from Norwegian waters from depths ranging from 36.3 to 818 m (1,200 to 2,700 ft). The report of this work was published in English as "On some remarkable forms of animal life from the great depths of the Norway coast," and was well received.

Almost all of Michael Sars's 45 papers on life cycles and larval transformations exist only in Norwegian, having been published by Oslo University. His reputation was such that his encouragement of the HMS *Challenger* expedition carried considerable weight. After much urging, the admiralty sponsored a globe-encircling, three and a half-year-long scientific expedition to study the ocean.

Sars received honorary doctorates from the universities of Zurich (1846) and Berlin (1860). He was a member of more than 20 academies of science and scientific societies. He died in Christiania on October 22, 1869, while planning other work.

When Sars began his zoological career, the animals he studied were considered a small group of simple, insignificant creatures. That notion, and the idea of the azoic zone, the region in the ocean where there was no life, were both put aside because of his work. The volume of studies on embryology was impressive and provided a firm foundation for later researchers both in Europe and the United States. The authority with which he spoke was recognized in his own time. His recommendation to authorize the expenditures on the *Challenger* voyage lent great weight to the importance of the plan.

⊠ Schafer, Wilhelm
(1912–1981)
German
Biologist, Paleontologist

Wilhelm Schafer was born in Crumstadt on March 18, 1912, son of a local pastor. After schooling in Mainz, he entered the university in Giessen for zoology, botany, and geography. He completed the requirements for the doctorate in zoology in 1937 with a dissertation on the development and color changes in the iridescent cells of *Sepia officinalis* (a relative of squid). His first professional appointment was at the Senckenberg Museum in Frankfurt, and his first administrative post there was as director of the branch of the museum in Wilhelmshaven. Schafer married Elizabeth Gotze, another zoologist, in 1939.

Schafer's research in Wilhelmshaven was not related to the topic of his Ph.D. He began an extensive study of sediment formation and of how both hard-shelled organisms and those with no shells might leave traces of themselves in the fossil record. He published several papers on sediment formation between 1938 and 1943, before he was drafted to serve in World War II. On returning to Wilhelmshaven in 1947, he found the museum a total wreck. As an industrial center, the city had been heavily bombed. Schafer spent years rebuilding it while continuing his research, and by 1954, he had published several other papers on *Brachyura*. This work was sufficient for him to qualify as a professor, and he was appointed to the zoology department at the university in Frankfurt.

His connection with the Senckenberg Museum continued. The main branch in Frankfurt was war-damaged, and Schafer remodeled it in 1962 to emphasize the fossil collection. In 1963, he published the results of much research on shallow-water sedimentation in the North Sea, stressing its relation to the fossil sediments. The next year saw the foundation of another branch of the Senckenberg, both as museum and research station, in Ischia, a city on the Gulf of Naples. In 1964–65, Schafer had his long-awaited chance for a long sea voyage. He sailed on the *Meteor* expedition that explored the coral reefs in the Red Sea. This voyage also found an interesting ecosystem that can withstand extremely hot water. Upon retirement in 1978, Schafer became interested in ecological and environmental concerns. He worked with several groups that were concerned about industrial pollution in and around the Rhine River and proposed remedial processes for both the river and its adjacent lands. Schafer's last book, *Fossilen Bilder und Gedanken auf Paleontologischen Wissenshaft* (Fossil images and thoughts on paleontological knowledge), published in Frankfurt in 1980, dealt with the centrality of paleontology in evolutionary history and philosophy, and the concept of man as a part of all nature. Schafer died in Frankfurt on August 27, 1981.

Schafer not only studied particular animals, he looked at them in the context of evolution according to the fossil record. In addition, he was a museum director, and as such, was a teacher. Good museums do not just illustrate; they instruct the viewers about the exhibits. Schafer wanted to add to the world's store of knowledge, and he wanted to share it with others.

⌧ Schmidt, Ernst Johannes
(1877–1933)
Danish
Biologist

Ernst Schmidt was born in Copenhagen, Denmark, on January 2, 1877, the son of Ernst and Camilla (Kjeldahl) Schmidt. His father died when the younger Ernst was a child, but the struggling family did manage to send the boy to school and then to the University of Copenhagen. His original field of study was botany, and he earned an M.S. degree in 1898 and a Ph.D. in 1903, both in botany. His doctoral dissertation was on mangrove prop roots. Schmidt's first position was in the Botanical Institute of the university, where he worked from 1899 to 1909.

Although Schmidt was a member of the Botanical Institute, his research interests were no longer directed exclusively toward plants. In 1899, he became a member of the Danish Commission for the Investigation of the Sea. From that point, he was increasingly interested in marine biology. Schmidt published a paper in 1904 that detailed the physiology of the larval stages of fish, particularly eels. The life cycle of eels was previously totally unknown. While these fish are plentiful in European waters, no one had ever seen a juvenile. Schmidt devoted the years 1908 to 1910 to the study of eels of the Mediterranean and North Atlantic. The research proved that eels in these two areas are not different animals but one species, and they breed in the Atlantic. This was surprising news.

In 1903, Schmidt married Ingeborg Jacobsen, daughter of the owner of the Carlsberg Breweries. This company, known for its philanthropy, found a cause close to its corporate heart, and in 1910, created the Carlsberg Physiological Laboratory at the University of Copenhagen. Ernst Schmidt was the director. He continued his research on eels with expeditions in 1920–22 to prove the paths of eel migration. The American eels, which are of a species different from that of European eels, were also shown to migrate. They, too, breed in the North Atlantic, and in roughly the same areas that are the breeding grounds for the European eels. The two species recognize and only breed with their own kind. This work was published in *Philosophical Transactions of the Royal Society of London* in 1922. A great deal of Schmidt's work was underwritten by the Carlsberg Foundation, which supported research on fish and on continental drift from 1928 to 1930. This was applied research: the major portion of Schmidt's work dealt directly with the concerns of commercial fisheries.

For his work, Schmidt was honored by a number of academies of science and other international organizations. He was an influential member of the International Permanent Council for the Study of the Sea, a group based in Copenhagen, and was still involved in its concerns when he died on February 21, 1933.

Schmidt's work on eel migration touches on two areas that have great interest for today's biologists. One is yet another facet in the complex field of species recognition. In every ecosystem, there are similar animals; how they locate and identify potential mates is an interesting research topic. Another part of eel migration involves a single animal living in both salt and fresh water. The physiological systems that must adapt and the way they do it are also of interest to current research. Schmidt's work provides a foundation for both.

⊠ Scoresby, William
(1789–1857)
British
Physicist, Navigator

William Scoresby was born near Whitby in northeastern England on October 5, 1789. His father, William, had run away to sea as a boy, and after considerable adventure and sailing experience in the North Atlantic, he came home to marry Mary Smith, a farmer's daughter. The elder Scoresby was a whaler—Whitby was then a whaling port—and in 1806, he tried and failed to reach latitude 89° North, for which the Admiralty had established a large prize. He did reach latitude 81° N. In the same year, the younger William Scoresby left Whitby for the university in Edinburgh, where his interest in science soon became apparent.

Scoresby left the university without a degree. He returned to Whitby, sailed with his father, and was for a time in the Royal Navy. There was regular shipping between Whitby and London, and on one of his trips there, Scoresby met Sir Joseph Banks, the famous naturalist who had ac-

companied JAMES COOK on Cook's second Pacific voyage. After that meeting, Scoresby returned to Edinburgh and finished his degree in 1811. He then returned to Whitby and married Eliza Lockwood the same year. By 1812, he was sailing as commander of the *Resolution*, and in 1813 commanded the *Esk*, a ship built in Whitby for both science and commerce.

On the *Esk*, Scoresby determined that the temperature of the sea below the surface was warmer than the surface water. He sent his findings to Joseph Banks, who arranged for the construction of precise thermometers. They proved unsuccessful since the wood casings swelled in the sea and broke the glass at depths of 300 fathoms (54.5 m or 1,800 ft). Scoresby redesigned the thermometers, and had the casings cast in brass and equipped with valves to withstand the increased water pressure. He called this heavy instrument the "Marine Diver," and in June 1817, it was lowered to a depth of 2,182 m (7,200 ft) in Arctic waters. This trial established the fact that the bottom water was much warmer than the surface and was an important factor in explaining the wealth of Arctic sea life.

Surveying Jan Mayen Island, Scoresby took rock and wildlife samples, as well as establishing that the recorded latitude and longitude of the island were incorrect. A decrease in the normal amount of polar ice made observation of the east coast of Greenland possible. On his return to England, Scoresby moved his family to Liverpool, where a new ship was built to his specifications. While waiting for his ship, the *Baffin*, to be built, he wrote his *Account of the Arctic Regions and Northern Whale Fishery*, a two-volume work that was published in 1820. This was considered the definitive work on high-latitude science. He went back to sea in the *Baffin* that year and again in 1822, when he successfully charted the eastern coast of Greenland. While he was away, his wife died, leaving him with a family to rear alone. One more voyage in 1823 yielded few new findings but did result in another book, published that year, which was a journal of his voyage to the Greenland area. Scoresby left the sea and joined the church, studied theology, and was ordained as a minister in 1825.

Maintaining his interests in marine science while practicing as a clergyman, Scoresby was asked in 1838 to design precise magnetic needles. The ones he submitted were not accepted, and after much argument with the Admiralty's using one of his specifications, he dropped the topic but retained interest in the subject. He wrote about his work in the field of geomagnetism in the two-volume *Magnetical Observations*, published serially between 1839 and 1852, and the account of an Arctic voyage in *The Franklin Expedition*, published in 1850. He traveled widely, to the United States twice, and to Australia in 1856, again to study magnetic properties of the Earth.

Scoresby contributed much practical information about the Arctic that formed a solid base for later scientists. Recognized as a competent scientist, he was elected a Fellow of both the Royal Society of Edinburgh and the Royal Society of London. William Scoresby died in Torquay on March 21, 1857.

Mary Scranton *(Courtesy Mary Scranton)*

⊠ Scranton, Mary Isabelle
(1950–   )
American
Oceanographer

Mary Scranton was born in Atlanta, Georgia, on February 28, 1950. Upon graduation from Mount Holyoke College in 1972, she began doctoral studies at the Massachusetts Institute of Technology and finished her work for a Ph.D. in oceanography in 1977. Scranton continued her research career at the U.S. Naval Research Laboratories. She is a member of the National Academy of Sciences and a research associate of the National Resource Council. Her present position is professor of oceanography in the State University of New York at Stony Brook.

Scranton's research activity concerns gases in the marine environment. Her studies explore the sources of gases in the ocean, what produces them, and what uses them, and the interaction of the chemicals in seawater with the biological processes of the organisms living in that water.

Scranton continues to produce research on the subjects of her specialty. Her recent papers discuss the fate of particular classes of compounds in restricted bodies of water, such as a bay. She is interested in water that maintains anoxic (oxygen-poor) communities. Certain compounds are recycled in particular ecosystems. Following the fate of these materials through the organisms in a community is painstaking, time-consuming work, which eventually yields information about the construction of the entire community. When one is devoted to the ongoing exploration of the interactions of many organisms, such as those that live in a coral reef, all of the organisms in and on the reef have to be studied in relation to the overall marine environment to understand the whole picture of how they affect each other.

⊠ **Shtokman, Vladimir Boresovich**
(1909–1968)
Russian
Oceanographer

Vladimir Shtokman was born on March 10, 1909, in Moscow. He entered the prestigious Faculty of Physics and Mathematics of the University of Moscow in 1928 to pursue a course of study in geophysics but suspended his academic career in 1932. At that point, he remained in the university as a laboratory assistant in the Institute of Oceanography. In the same year, after promotion to the staff, he published his first work on the study of undersea cables. Every country was studying undersea cables because they were a necessity and a problem. They broke when there was no detectable cause, disrupting communication on a global scale. Other scientists were examining the same problem, and it remained insoluble until after World War II. Shtokman was promoted for his efforts, and in 1933, named leader of an expedition to the Barents Sea.

In 1934, changing from the study of marine conditions to those of an inland sea, Shtokman moved from Moscow to Baku, Azerbaijan, where he created a laboratory for physical oceanography as part of the Soviet Union's All-Union Scientific Research Institute of Ocean Fisheries and Oceanography (UNIRO). His research interest was the mathematical study of ocean turbulence; this was pioneering work in the USSR. Shtokman collected data for this study in the Caspian Sea, where he had led expeditions that collected data explaining turbulent pulsations. While in Baku, Shtokman finished his work for a Ph.D. in physics and mathematics in 1938. He began his

investigations into the effect of wind stress while there and found that in an enclosed sea it produced countercurrents. This type of current also exists in open ocean, particularly in the tropics.

During the early 1940s Shtokman was director of UNIRO and on the staff at the Institute of Theoretical Geophysics of the USSR Academy of Science. This institution was based in Krasnoyarsk, which was relatively unaffected by the war. He studied Greenland Sea currents and the patterns of Atlantic water flow into the Arctic. In 1943, Shtokman returned to Moscow to defend his dissertation on Arctic waters. Once there, he created a laboratory for oceanography for the Soviet Academy of Science, where he spent the rest of his life as director.

Using VAGN EKMAN, the Swedish physicist and theoretician, he explained mathematically what was entailed in "right-hand-drift"; layers of air and water spiralled to the right (in the northern hemisphere) with the lowest layer the most displaced. These layers of water can be envisioned as steps of a spiral stair as seen from above. Each step is slightly to the right of the one above it. Shtokman, the major figure in Soviet oceanography and ocean currents, demonstrated the stress of wind on water and the results created by that wind, showing that the tangential wind stress on a body of open water contributes significantly to the horizontal circulation in the ocean. This important work appeared in 1945. He continued the study of wind stress, publishing in 1946 on the theoretical methodology for determining velocities of winds in relation to the depth of the water they move over. The driving force for this work was a Soviet desire to simplify the geographic plotting of the waters and currents surrounding the country. Shtokman's study was the basis for the work of HARALD SVERDRUP on the modeling of ocean currents. Two years later, Shtokman published another significant work showing that many irregularities of the equatorial countercurrents are the result of the trade winds in lower latitudes.

Shtokman's last major work described the circulation of water around large islands in a direction opposite to that of the circulation in the surrounding ocean. This phenomenon is observed in Iceland, the Kuril Islands, and Taiwan. In 1966, he published his explanation of this observation; the bulk of the island redirects the wind, and the wind stress interferes with the horizontal movement of the water. Shtokman died suddenly, in Moscow, on June 14, 1968.

⊠ **Smith, Deborah K.**
(1950–)
American
Oceanographer

Deborah Smith received her first undergraduate degree in mathematics in 1972 from the University of Wisconsin. In 1981, she received a B.S. in geology from San Francisco State University, and in 1985, still on the West Coast, she received a Ph.D. in earth sciences from the Scripps Institution of Oceanography, University of California, San Diego. Her dissertation was "Seamount Populations in the Pacific Ocean." She stayed at Scripps as a research assistant for a year and then was a post-doctoral fellow at Woods Hole Oceanographic Institution (WHOI) in Massachusetts. Smith is now a senior scientist there (a rank at that research institution equivalent to professor). Her research interests are the variability of the topography of the sea bottom, statistical modeling of the seafloor, seafloor formation, seamounts, and modeling of undersea volcanic activity. Smith and coworkers explore an active work site, the Puna Ridge (the undersea ridge near the Hawaiian Islands).

Smith has won a number of awards. The American Association of Petroleum Geologists (1981), the Office of Naval Research (1987), and from the Mellon Independent Study Fund (1992, 1995, and 2001) have all honored her. She was a visiting professor at the Laboratoire de pétrologie,

Université Pierre et Marie Curie, in Paris in 2000–01. She is a member of the U.S. Science Advisory Committee, the American Geophysical Union, and the Geological Society of America, and was the editor of the *Journal of Geophysical Research (Solid Earth)* from 1988 through 1994. Smith's work appears in important geological journals such as *Journal of Volcanism and Geothermal Research* and *Geophysical Research Letters*.

She is a primary researcher working for a prestigious institution and actively engaged in studying the geology of a rapidly changing part of the planet's surface. Volcanoes and earthquake activity on land are obvious. The same processes occur with much greater frequency undersea, and that makes them both interesting and difficult to study. These processes affect life on land, since the weather, sea life, and its populations are affected, and tsunamis that totally disrupt coastlines may be created.

In addition to an active career in fundamental research, in 1998 Smith designed and developed a website (http://www.punaridge.org) and the teaching materials to go with it to give schoolchildren a "virtual seagoing research expedition" experience. Another website that Smith aimed at schoolchildren is one that brings them into the research world of women oceanographers (http://www.womenoceanographers.org).

⊠ **Smith, Sidney Irving**
(1843–1926)
American
Zoologist

Born in Norway, Maine, on February 18, 1843, to Elliot and Lavinia (Barton) Smith, Sidney Smith was a seventh-generation New Englander. His interest in science appeared early. As a boy, he collected insects. His assemblage and classification so impressed LOUIS AGASSIZ that Agassiz bought his collection for Harvard College. Surprisingly, he did not study with Agassiz; instead, in 1864,

Smith went to the Sheffield Scientific School, attached to Yale. Yale, like Harvard, did not have an undergraduate facility for science; Sheffield was a separate, albeit connected, school. Working with ADDISON VERRILL, Smith completed the equivalent of a B.S. degree in 1867 and moved into the position of assistant in zoology at Sheffield.

As an undergraduate, Smith had spent summers dredging and collecting crustacean specimens, and he continued that work while an assistant. His voyages with Verrill to Long Island Sound and to the Bay of Fundy (on the east coast of Canada) occupied the summers from 1867 through 1870. Leaving Yale, he then spent a year as the zoologist for the U.S. Lake Survey at Lake Superior. This was the inland analog of the U.S. Coast Survey and just as wide-ranging in its interest in the total environment of the body of water being studied.

In 1872, Smith moved to the U.S. Coast Survey and worked on an area in the St. Georges Bank off the North American coast. Joining the U.S. Fish Commission was a natural next step for Smith. "Metamorphosis of the Invertebrate Animals . . ." in *Report of the U.S. Fish and Fisheries Commission*, 1871–72, and "Early Stages of the American Lobster," in *Transactions of the Connecticut Academy of Arts and Sciences*, 1873, are two results of Smith's work at the fish commission. By 1875, Yale University recognized that science was an increasing presence in the university and created a professorship in comparative anatomy within the zoology department. Yale offered the post to Sidney Smith and he accepted. He remained in this position until his retirement in 1906.

A long voyage of exploration and dredging in 1876 took Smith to the Kerguelen Archipelago in the southern Indian Ocean off the Antarctic coast. Two fish commission vessels, the *Fish Hawk* and the *Albatross*, were made available to Smith and his coworkers for Pacific dredging in 1880 and 1883 respectively. He collected crustaceans on

both voyages and using them as analogs of the creatures he knew from similar collecting trips in Atlantic waters. The latter voyage marked the end of Smith's career as a sea-going researcher. Smith continued to teach; he loved it, and was an effective instructor in biology at several levels. Since he was interested in the science education of medical students, he created the first biology course for this group at Yale. It was a pioneering effort in medical education. He was also active in the organization of the Woods Hole Oceanographic Institution, Massachusetts, which came to fruition in 1930.

As a child, Smith collected and classified insects and his interest in entomology was a high-level hobby throughout his working career at Yale. In his post-retirement years, Smith was the state entomologist for both Connecticut and Maine. He was still an entomologist when he died in New Haven, Connecticut, on May 6, 1926. His large collection of insect specimens was willed to the Peabody Museum of Natural History at Yale and the National Museum of Natural History in Washington, D.C.

Suess, Eduard
(1831–1914)
Austrian
Geologist

Eduard Suess was born in London, England, on August 20, 1831; his father's wool business had taken the whole family there. They returned to Vienna in 1834 and spoke English in the household. As a result, the young Eduard was always bilingual, as well as a gifted student. He entered the gymnasium (secondary school) early and was part of the revolutionary stirrings in the school population. Fearing that he had tuberculosis, the family sent him to live with his grandparents in Prague; he collected mineral specimens and attend the University of Prague. While there, he missed the revolutionary upheavals that beset Vienna in 1848 and found his calling: geology.

Edward Suess *(Hulton/Archive/Getty Images)*

A manuscript on geology that he sent to the Viennese Friends of Science and his friendship with the paleontologist Joachim Barrande led to a position for Suess in the Museum of Natural History. There he was the museum representative to the foreign correspondents. This fairly insignificant public relations position brought the charming young man into contact with many people. Later, Suess's trip to the spa at Carlsbad (now Karlovy Vary in the Czech Republic) led him to make comparisons of the rock formations there with the rocks around Prague and Vienna. This resulted in a chapter on geology in a tourist guidebook, which was Suess's first published work.

Marriage to Hermine Strauss furthered Suess's career in geology; she was the daughter of a socially prominent doctor and the niece of the director of the Museum of Natural History. When

Suess applied in 1856 for the post of *Privatdozent* (research position in geology) at the University of Vienna, he was turned down because he had no doctorate. However, Wilhelm Hardinger, who was a member of the Viennese Friends of Science and also the minister of education, intervened, and Suess became *Privatdozent* and eventually professor of geology. He continued his practice of going on specimen-collecting tours with other geologists; for example, in 1854, he went with a party to Switzerland; in 1856, he traveled to France and Germany; in 1862, he went to England; and in 1867, he explored rock formations in Italy.

Working within the small circle of people involved in science and in the university, Suess was active in the promotion of science education in popular universities that were open to a wide range of society. Embarking on a political career, he served on Vienna's city council from 1863 to 1886, in the provincial diet (assembly) from 1869 to 1896, and in the national parliament from 1872 to 1896.

Suess's writings include some that pertain to his political work, such as the book *Die Boden der Stat Wien* (The walls of Vienna), published in 1862, written about the removal of the city walls in 1857. This modernization and expansion of the city changed it from a medieval settlement on a boggy stretch of the Danube to a modern city, and had a profound effect on the direction of the Danube River and the water distribution in the city. Before this extensive building project was undertaken, there was frequent flooding in Vienna. The floods were devastating events because the water supply became contaminated, and there were many deaths from typhoid fever. As a politician, Suess urged the construction of an aqueduct (built in 1873) and the Danube Canal (opened in 1875). These civic building projects totally eliminated the threat of flooding and greatly improved living conditions in Vienna.

His scientific writings include some that went back to his original interest in paleontology; he studied graptolites, brachiopods, ammonites, and

Tertiary mammals. Graptolites are extinct animals that lived in a colony looking like a tree. They are possibly related to Hemichordata (animals with a dorsal nervous system that is not necessarily encased in bone). Graptolites are frequently found near brachiopods, but if there is a connection between them, it is unknown. Brachiopods are bivalves, mollusks that superficially look like clams but have a totally different lineage. Ammonites are a large, extinct group of shelled animals whose shells grew larger by adding another segment. They were very successful and widespread for a long geologic time and are used to date the layers in which their fossils are found. The closest extant organism related to them is the chambered nautilus found in the Indian Ocean. These studies may place Suess as the first paleobiologist.

Suess's thoughts on geology are the reason for his fame. He went from a "unity of the living world" concept in 1875—the last remains of romanticism—to the idea of a "biosphere"—a modern concept of the Earth as a whole, closed system. His acceptance of the then current theory of geosynclines as an explanation for the shapes of rock in the Earth changed. Geosyncline was the name created to explain a trough cutting through an established series of layers of rock before the presently accepted theory of crustal plates floating on a semi-liquid mantle. Now we would call this a plate edge. His own observations led him to theorize tectonics; possibly because he had had no formal training in paleontology or geology. He was self-taught and free of the accepted theories then current.

Given a commission in 1865, Suess was funded to write an analysis of the geology of the Austrian Empire. He worked on it for years before he deemed it ready for publication. The result was *Die Enstehung der Alpen* (The origin of the Alps), published in Vienna in 1875. This was more than a description of the Alps and their effect on the empire; it was the mechanics of mountain chain formation as a global process. His next work was his masterpiece, *Das Anlitz der*

Erde (The face of the earth). This three-volume work, published in Vienna, appeared in sections between 1883 and 1909. It was translated into French as *La face de la terre*, in four volumes, published in Paris between 1897 and 1918; and in a five-volume English edition, published in Oxford between 1904 and 1924.

This work was the definitive geology text. In it, Suess synthesized many thoughts by a number of earlier geologists, geographers, and paleontologists. He theorized that one primordial landmass, Pangaea, had broken up into five proto-continents: Laurentia, Fennoscandia (a term invented by Sir William Ramsay), Gondwana, Angara, and Antarctica. He also envisioned a vast ocean, the Tethys Sea, the remnants of which are the Mediterranean, Aral, Caspian, and Azov Seas, and documented how the ancient seawater had formed shorelines and changes. The work extrapolated seafloor alterations from events on land. Seacoasts rose and then fell, and rose again, leaving the central massifs intact. According to Suess, volcanoes blew gas out of the Earth, and this later condensed to form ocean water. This provocative work influenced many other researchers. One directly affected was Suess's student, ALFRED WEGENER. Their later efforts produced what historians of science called a "revolution in science," plate tectonic theory. On retirement, Suess moved to Marz in Burgenland, Austria, where he died on April 26, 1914.

Sverdrup, Harald Ulrik
(1888–1957)
Norwegian
Oceanographer

Harald Sverdrup was born in Sogndal, Norway, on November 15, 1888, to Johan and Varia (Vollan) Sverdrup. His father and four uncles were ministers in the Norwegian State Church (Lutheran). The family later moved to an island district, Solund, north of Bergen. Since his mother had died when Harald was a child, and the family moved frequently, he was raised by governesses. At 14, he was sent to *gymnasium* (secondary school) in Stavanger. At that time, he was beginning to be interested in science and felt that it was irreconcilable with his family's interests. He therefore did not choose the science curriculum. By the time Sverdrup was finishing his career at *gymnasium* he had realized that the university and theology were not synonymous, but he spent the year (1907–08) before entering the university at the Norwegian Academy of War in compulsory military service. He emerged as a reserve officer educated in physics and mathematics, and the endurance training received there probably made it possible for him to survive some of his Arctic travels.

Harald Sverdrup *(American Geophysical Union, courtesy AIP Emilio Segrè Visual Archives)*

Entering the University of Christiania in 1908, Sverdrup thought that he was going to be an astronomer. Since the curriculum of physical geography and astronomy was sufficiently broad to include what is now geophysics, meteorology, oceanography, and geomagnetism, he received an extensive education. His path was determined when VILHELM BJERKNES, the most prestigious contemporary scientist born in Norway and internationally known as a meteorologist, offered him an assistantship. Bjerknes's research group was based in Bergen, where Sverdrup was one of the best and brightest young Norwegian scientists. Although his interest in astronomy continued, meteorology and oceanography were becoming more fascinating. When Bjerknes moved his research to the new Geophysical Institute in Leipzig, Germany, in January 1913, Sverdrup went with him. While in Leipzig, Sverdrup finished his thesis on North Atlantic trade winds for a doctorate from the University of Christiania. The degree was granted in June 1917, and Bjerknes and his party returned to Norway the following August.

On July 18, 1918, Sverdrup accompanied Roald Amundsen on an expedition to the North Pole on the *Maud*, a research vessel. Amundsen had asked Sverdrup to do this once before, in 1913, but he had declined, since it would have interrupted his university studies. The trip was projected as one of three to four years' duration, but it took nearly twice that long. They finally returned on December 22, 1925, after a 10-month hiatus (1921–22) when Sverdrup was in the United States visiting the Carnegie Institution in Washington, D.C. There, he analyzed the electric and geomagnetic findings of the *Maud* expedition. For a period in 1919–20 when the vessel was ice-bound, Sverdrup lived with the Chukchi, nomadic Siberian reindeer hunters. Two works that resulted from his expedition with Amundsen are "Meteorology on Captain Amundsen's Present Arctic Expedition" (1922) and "The North-Polar Cover of Cold Air" (1925), both published in the *Monthly Weather Review*.

By 1926, Sverdrup was established as a scientist. He was then offered the chair in meteorology at Bergen and accepted it. His research activity was organizing and writing the report of the *Maud* expedition. In 1930, he was back in Washington, analyzing new data retrieved by the research vessel *Carnegie*, and he published "Some Oceanographic Results of the *Carnegie*'s Work in the Pacific—the Peruvian Current," in *Transactions of the American Geophysical Union* (1930), based on that work. A new position as research professor at the newly organized Christian Michelsens Institute was created for Sverdrup in 1931. He also led the scientific group in the Wilkins-Ellsworth North Polar Submarine Expedition. Their ship, *Nautilus*, did not reach the pole but did collect considerable data on Arctic currents. In 1934, a new two-month project had Sverdrup investigating boundary layers in snowfields in Spitsbergen, a Norwegian archipelago in Arctic waters.

A major change in Sverdrup's life occurred late in 1935. BJØRN HELLAND-HANSEN, the director of the Christian Michelsens Institute, returned from the United States and reported that Sverdrup had been one of the choices for director of the Scripps Institution of Oceanography (SIO) in La Jolla, California. He rather reluctantly accepted the post when it was finally offered, stipulating that it would be for three years. At the time of Sverdrup's arrival in August 1936, SIO was tiny; it had one coastal vessel, eight faculty members, and a total staff of 30. But it was about to become a world-class research installation.

His first task was to organize the faculty and concentrate their efforts on a single project: a hydrographic study of the California Current was the first result of this program. The next task for Sverdrup was to foster close ties to the University of California at Los Angeles (UCLA). Scripps was then a research institution. It did not grant degrees but used its connections to the University of California, which gave degrees to trained researchers for work done at Scripps. UCLA was

also the university where fellow Norwegians JAKOB BJERKNES and Jorgen Holmboe were developing a meteorology department. The association with UCLA was important in the reform of the oceanography curriculum. A realignment of the faculty produced a department of oceanography that concentrated the skills and talents of the personnel on the geology, physics, and chemistry of oceans.

Sverdrup was fortunate in his staff, which included ROGER REVELLE, and then the brilliant student and later staff member WALTER MUNK. Work on a text for oceanography was another result of the curriculum reform. With fellow faculty coauthors Martin Johnson and Richard Fleming, Sverdrup produced *The Oceans: Their Physics, Chemistry, and General Biology* in 1942. Its military implications were so important that its foreign distribution was forbidden.

The three-year term Sverdrup had requested as director of Scripps ended in 1939, when World War II began in Europe. Realizing that his stay in La Jolla would be indefinite, Sverdrup became an American citizen. This eliminated some but not all of his problems with security clearance; he was from an occupied country where many of his close relatives still lived. In spite of those complications, he and Munk worked on wave forecasting, a topic of vital interest to the invasion planning of the military. All of SIO's research efforts were redirected to wartime needs early in 1942. By 1943, Sverdrup was planning for SIO's postwar future. This included much help from the U.S. Navy since the University of California was convinced that once the war ended, the need for an oceanographic institute would be gone.

In a concerted effort to promote support for scientific research, Sverdrup became an active figure on a number of scientific committees. He was elected to the National Academy of Sciences in 1945, to the Executive Committee of the American Geophysical Union (AGU) in 1945, and to the presidency of the AGU's Oceanography Section. The next year, he became president of the International Association of Physical Oceanography and also was chairman of the Division of Oceanography and Meteorology in the Pacific Sciences Conference. He was also instrumental in the establishment of an Institute of Geophysics and Planetary Physics, a long-term study of marine fisheries, and the acquisition of new oceangoing research vessels.

By 1948 Sverdrup was ready to return to Norway. Once there he organized the Norwegian Polar Institute and served as its director. This organization launched a three-nation (Norway, Great Britain, and Sweden) expedition to Antarctica in 1949–52. Sverdrup remained in Oslo (the new name of Christiania adopted in 1925), where he accepted a post as professor of geophysics at the university in 1949. He also served as dean of the Faculty of Science and vice director of the university. One of his achievements as an administrator was the committee work resulting in the reorganization of the Norwegian educational system. Harald Sverdrup died suddenly on August 21, 1957. A heart problem that he had decided not to pamper had resurfaced. His voluminous correspondence and papers are archived at the Norwegian Polar Institute in Oslo.

T

Tennent, David Hilt
(1873–1941)
American
Biologist

A native of Janesville, Wisconsin, David Tennent was born to Thomas and Mary Tennent on May 28, 1873. He had an uneventful childhood and was planning to join his father in a medical career. When his father died, Tennent gave up his medical studies and became a pharmacist instead, to help support his family. At the time, a pharmacy license was granted if one passed the licensing examination. Tennent taught himself what he needed to know using the official publication, the "Dispensatory of the United States of America," and the experience he had while working in a pharmacy as a high school student. However, before he had a chance to practice pharmacology, one of his older sisters helped Tennent to enroll in the science program at Olivet College in Michigan in 1895. His study for the degree in biology was longer than most because he was working at the same time, but he received a B.S. in 1900 and was strongly encouraged to continue his studies. A scholarship to Johns Hopkins University in Baltimore, Maryland, was made available with the help of his biology teacher at Olivet, Hubert Lyman Clark. Clark later became the curator of marine invertebrates at Harvard's Museum of Comparative Zoology. At Johns Hopkins, Tennent spent the summers of his student years at the fisheries laboratory in Beaufort, North Carolina, where the whole Johns Hopkins group collected to study the marine plants and animals of that part of the Atlantic coastline. His adviser for the Ph.D., granted in 1904, was William Brookes. His doctoral dissertation was "A Study of the Life History of *Bucephalus haimeanus:* A Parasite of the Oyster."

Tennent's first appointment proved to be the only one; he became an instructor in the biology department at Bryn Mawr College in 1904, married Margaret Maddux, a Bryn Mawr alumna, in 1909, and rose to be professor in the same department. He spent his summers in the field collecting specimens. He spent several summers in the Dry Tortugas at the Biological Laboratory of the Carnegie Institution. The first was in 1909. "The Chromosomes in Cross-Fertilized Eggs," published in *Biological Bulletin* in 1907, and "Experiments in Echinoderm Hybridization," in the *Carnegie Institution (Washington) Yearbook* in 1909, were two publications arising from this work. Tennent continued collecting in 1911 at the Carnegie installation in Jamaica, British West Indies; at Cold Spring Harbor on Long Island (New York); at the zoological station in Naples, Italy; at the Hopkins marine station in Pacific Grove, California; and the Marine Biological Laboratory in Woods Hole, Massachusetts.

From 1937 to 1940, when it closed, Tennent was the executive officer of the laboratory.

A long leave for a sea voyage in 1913 took Tennent on a sampling expedition to the Torres Strait (between New Guinea and Australia), and on a yearlong sabbatical leave, he went to the laboratory of the Imperial University of Tokyo at Misaki (1922–23). At the end of his stay, he deposited his year's work in a bank vault and went on vacation in China. While he was away, an earthquake destroyed the bank and his year's work. Some of this material was reconstructed from notes and published. Tennent returned to Japan for another sabbatical year at Keio University in Tokyo in 1929–30.

The work that David Tennent is known for concerns echinoderms specifically, but ultimately has significance in all egg fertilization and development. He began these areas of research as a graduate student. His first experiments were with the species cross-fertilization of echinoderm eggs to produce hybrids. Cross-species fertilization can happen if the external environment is manipulated, for instance, by increasing salt concentration, but it is unusual, and the hybrids are almost always sterile. When he began his research, Gregor Mendel's work explaining inheritance had been rediscovered. Mendel had formulated the laws of inheritance, which explained how a trait could be present in the genetic material of an organism without being apparent, and reappear unchanged in a descendant. This was basic information for a researcher who was looking into the nature of inheritance. Tennent was tracing the lineage of both the egg's and the sperm's chromosomal contribution to the final sea urchin embryo. He found that the sea urchin's sperm is digametic. The sex chromosomes (gametes) are different in the egg and in the sperm, and carry different traits. This discovery led to his careful study of which traits came from which parent.

Carrying this research further, Tennent varied the environment in which the eggs were developing, using an array of salts to change the nature of the seawater: $NaCl$ (sodium chloride), $BaCl_2$ (barium chloride), $CaCl_2$ (calcium chloride), and $SrCl_2$ (strontium chloride). Assuming that the sperm of different species differs in its effect on the egg, the exposure of the egg to one or another of the solutions of seawater to which the salts were added would affect the likelihood of hybridization. The sodium ion increased the permeability of egg cell membranes; other ions decreased it. The end result of these experiments was the proof that sperm acted preferentially on an egg of the same species but that chances for hybridization were enhanced by the change in the ionic character of the surrounding water. This research was published as "Evidence on the Nature of Nuclear Activity," in *Proceedings of the National Academy of Science* in 1920. Tennent's work gained importance for later generations of biologists, when they in turn worked on the nature of cell recognition. Tennent also concentrated on the specific structures of the developing fertilized egg to determine which organs grew from specific parts of the gastrula, an early stage in the developing embryo. It occurs after the embryo's cells form a hollow ball, which collapses into what looks like a deeply creased, deflated ball. That is the gastrula stage, and it is crucial in the development of the organism. This is the point from which cell differentiation occurs.

The last research project that Tennent worked on concerned the nature of the developmental response to light. He tested this using a dye that was nontoxic to the developing sea urchin embryo in darkness but became toxic in light. "The Photodynamic Effects of Vital Dyes on Fertilized Sea Urchin Eggs," published in *Science* in 1935, and "Some Problems in the Study of Photosensitization" in *American Naturalist* in 1938, were the results of these projects.

Serving on university and local committees in Bryn Mawr was part of Tennent's life. Professionally, he also served as President of the American Society of Zoologists (1916) and the American Society of Naturalists (1937). He was elected to the national Academy of Sciences in 1929 and the American Philosophical Society in 1938. When he retired from teaching in 1938, he continued to work as a research professor until his sudden death on January 14, 1941.

⊠   Tharp, Marie
(1920–   )
American
Cartographer, Oceanographer

The child of a U.S. Department of Agriculture surveyor who made soil classification maps, Marie Tharp grew up in a milieu where cartography was a household word. Nonetheless, according to Marie Tharp, "I never would have gotten the chance to study geology if it hadn't been for Pearl Harbor." After completing a B.A. in English and music in 1943, she went on to the University of Michigan, which had recently made openings available to women due to the shortage of male students during World War II. Tharp completed an M.S. in geology in 1944 and an M.S. in mathematics in 1948, while working for an oil company in Tulsa, Oklahoma.

"Hooked on research," she came to New York in 1948 and found a position as an assistant to a graduate student, BRUCE HEEZEN. W. MAURICE EWING, the head of the department, and soon director, of the Lamont Geological Observatory (later known as the Lamont-Doherty

Marie Tharp in the early 1950s at the Lamont Geological Observatory. *(Courtesy Lamont-Doherty Earth Observatory)*

Earth Observatory) of Columbia University, had one request: "Can you draft?" Beginning in 1952, Tharp spent years piecing together the profiles of sections of ocean bottom based on the soundings taken by Ewing and Heezen. Each section was 1° of latitude by 1° of longitude. It soon became clear that the mountains of the 40,000-mile-long (64,000 km) mid-ocean ridge did not match up; there was an obvious cleft running down the center of the ridges with mountains on either side. This suggested to Tharp the likelihood of a rift valley, a steep cleft between two long, steep mountain ridges.

The idea was dismissed at first by Ewing and Heezen, but when frequent earthquakes occurred along the rift valley releasing quantities of molten rock, the idea gained credibility. The work done by Tharp and her map changed the way geologists thought about the Earth. Her work was an integral part of the evidence for the theory of plate tectonics, the idea that the seafloor is moving and spreading from the mid-ocean ridges outward. Her famous map, entitled "the World Ocean Floor," was published in 1977 by the National Geographic Society.

The Library of Congress honored Tharp in 1997 when she was noted as one of four people who had made major contributions to cartography, and again in 1998 as part of the 100th anniversary of the Library of Congress's Geography and Map division. She had been given awards for her contributions to oceanography by the Woods Hole Oceanographic Institution (1999) and by the Lamont-Doherty Earth Observatory of Columbia University (2001).

⊠ **Thompson, John Vaughan**
(1779–1847)
British
Zoologist

John Thompson was born in Berwick-upon-Tweed, England, on November 19, 1779. He be-

came a career military officer who maintained an intense interest in science. In 1799, Thompson was an assistant surgeon in the Prince of Wales Regiment and, after a promotion to surgeon, in 1835 was appointed medical officer in a convict settlement in New South Wales, Australia.

The research areas of John Thompson's interest fall into three groups. Botany was the first; while he served in the Napoleonic wars, Thompson spent a total of four years in various postings, including Gibraltar, the West Indies, and the Netherlands. Wherever he went, he first collected and preserved botanical specimens and then moved on to zoology in general and finally concentrated on marine zoology.

His second scientific endeavor involved microscopic marine invertebrates. On returning to Europe on leave from his posting in Australia in 1816, he saw a phosphorescent display in the sea off the island Mauritius in the Indian Ocean. This inspired him to identify the organisms that created the phenomenon. Thompson is thus credited as the first to use a plankton net. This is really a bag made of very finely woven silk. Another researcher, John Cranch, had also used such a net, at roughly the same time. Cranch, however, died shortly thereafter, and Thompson went on to use the net in other research.

In 1827, he published results from his third area of scientific interest, entitled "Memoir on the *Pentacrinus europaeus*." This is a shallow-water European crinoid, a sea lily. It is an echinoderm, a relative of starfish and sand dollars.

Up to that point, living sea lilies and feather stars were known only in the West Indies, not in Europe. In 1836, he published again on a commensal polychete worm living in the *P. europaeus*. A commensal relationship is one between two (or more) organisms, where each benefits from the association. A polychete is a segmented marine worm that usually has an attractive feathery array of tentacles around its mouth. This acts as a lure for organisms that are snapped up as food to be used by both the worm and the crinoid.

The structure of the crinoid provides the housing for the worm.

One particularly noteworthy discovery made by Thompson in the early 1830s was that the plankton first identified as *Zoea* were really the larvae of the common European crab, *Cancer pagurus*. The larva does not look at all like the adult, and so when it was found, it was named as a unique organism. This was a significant find. Since the classification of GEORGES CUVIER was still widely used and he was recognized as the highest authority on classification, considerable controversy attended Thompson's determination of the relationship between the two organisms. With similar work on the pea crab, Thompson identified that organism as a parasite living within the shells of mussels and spider crabs. His next important find concerned cirripeds, the barnacles. These organisms had been known for centuries. They are of commercial importance since they encrust and befoul moorings and ship bottoms, which, as a result, require constant scrapping and cleaning. Barnacles were assumed to be related to clams and limpets (other shelled mollusks) because, superficially, they look like mollusks. They are highly modified and complicated crustaceans. Twenty years later, these creatures received intense study from CHARLES DARWIN.

Again Thompson rebutted Cuvier's classification and courted controversy by naming a group of planktonic organisms *Polyzoa* and identifying them correctly as Bryozoa, neither hydroids nor ascideans. Hydroids are tubular coelenterates, such as coral organisms. Ascideans are free-living marine roundworms.

Thompson produced very few publications. One example is *Zoological Researches and Illustrations*, a set of five extremely rare, hand-colored pamphlets that appeared between 1828 and 1834. Other publications appeared in the *Zoological Journal* in 1831 and in *The Entomological Magazine* in 1835 and 1836. After retirement in 1847, Thompson remained in Australia and died in Sydney on January 21, 1847.

⊠ **Thomson, Sir Charles Wyville**
(1830–1882)
British
Marine Biologist, Oceanographer

Wyville Thomson, who most certainly thought of himself as a Scot, was born in Bonsyde House in Linlithgow, Scotland, on March 5, 1830. His father, Andrew Thomson, was a surgeon in the East India Company, therefore it was assumed that the son would follow the same path. After education at the Merchiston Castle School, the young Thomson enrolled in the University of Edinburgh's famous medical school at the age of 16. He became ill and, because of an interest in natural history, attended the lectures in zoology, botany, and geology. This led to a lecturer's position at the University of Aberdeen in 1851 and,

Wyville Thomson *(Hulton/Archive/Getty Images)*

after a year, to a position at the university's Marischal College.

Two years later, he moved to the post of professor of natural history at Queen's College in Cork, Ireland, and married Jane Ramage Dawson. Thomson stayed in Cork for only one year. By 1854, he had moved to Belfast, where he was professor of zoology and botany. While in Belfast, Thomson became known as Britain's most knowledgeable marine biologist, giving public lectures and writing of his collecting and classification of coelenterates, polyzoans, and fossil cirripeds (barnacles), trilobites (extinct crustacea) and crinoids (sea lilies). Coelenterates are a numerous phylum of organisms, some of which have tubular bodies typified by the coral organisms, others are mobile, umbrella-shaped masses of jelly, the medusas, commonly called jellyfish.

Thomson was part of the debate concerning the presence of living organisms in ocean depths. At the time, EDWARD FORBES's idea of an azoic zone—a region devoid of life—in the ocean depths was a credible theory. A trip in 1866 to visit MICHAEL SARS established Thomson as a deep-sea biologist, and he began to envision a scientific survey of the open ocean.

Important changes occurred in Thomson's life in 1868; he moved to a new post as professor of zoology and botany at the Royal College of Science in Dublin, and together with WILLIAM CARPENTER, then vice president of the Royal Society, he began to petition the society and the admiralty to support a deep-sea dredging voyage in the North Atlantic. Their first effort was on the HMS *Lightning* in 1868. The dredge brought up sponges, rhizopods, echinoderms (starfish), crustacea, mollusks, and foraminifera (elaborately shelled one-celled organisms) from depths greater than Forbes's limit of 300 fathoms (350 m or 1,200 ft). They also found variations in water temperature, dispelling another belief that the sea was all a uniform 4° C. The next year after these well-received findings, Thomson

continued dredging, chemical studies of water composition, and temperature tests aboard the HMS *Porcupine*. Living organisms were retrieved from a record depth of 2,435 fathoms (4,428 m, or 14,610 ft).

Thomson was elected a fellow of the Royal Society in 1869 and was funded for another exploratory cruise, this time to the Mediterranean in 1870. The same year he was appointed as Regius professor of natural history at the University of Edinburgh. There he finished his massive work, *The Depths of the Sea*, a detailed account of his years of research work, published in 1873.

After years of campaigning for such a grant, Thomson led the civilian scientists aboard the HMS *Challenger*, a steam corvette that sailed on December 7, 1872, for a world scientific cruise. It returned on May 24, 1876, having traveled 68,890 nautical miles (127,584.3 km) and made 362 soundings. The voyage cost over £100,000, then a colossal sum. Some of the expedition's unique finds were the discovery of living organisms at depths greater than 3,000 fathoms (5,455 m or 18,000 ft), and the presence of nodules of manganese pentoxide. JOHN MURRAY, the chief ship's assistant on the *Challenger*, collected ocean sediments and oozes. The clay bottom was unusual; at very great depth and pressure the shells of foraminifera dissolve, since their calcium carbonate becomes more soluble. Murray's work on the globigerina and radiolarian oozes had never been seen before, and his discovery of the diurnal migration of plankton created a sensation. There was no physicist aboard the *Challenger*, and therefore, questions about ocean circulation were left unanswered. A significant find was the vast undersea mountain range, the mid-ocean ridge. While there had previously been reports of irregularities on the ocean bottom, notably those made by MATTHEW MAURY, the findings brought back by the *Challenger* were startling. Although many found it implausible, it was confirmed on the voyage of the *Meteor*, led by GEORG WÜST in

1925–27. Upon his return, Thomson first wrote a preliminary report, *The Voyage of the Challenger*, published in 1876, the year he was knighted. He began editing the many monographs (there were eventually more than 50), detailing aspects of the voyage, but he became ill, and resigned in 1881. He died in Bonsyde House on March 10, 1882.

V

Vening Meinesz, Felix Andries
(1887–1966)
Dutch
Geophysicist

Felix Vening Meinesz (often referred to as Felix Meinesz) was born on July 30, 1887, in Scheveningen, a seaside resort not far from Rotterdam, where his father, Sjoerd Vening Meinesz was burgermeister (mayor). Felix was interested in engineering and studied that at the University of Delft, receiving a B.S.E. in 1910. Before graduation, he participated in a gravimetric study (a study of gravity and its differences at different places on Earth) of the Netherlands, an important subject for a nation where much of the arable land is below sea level. The Netherlands was a marshy area at the mouth of the Rhine and the Meuse Rivers. The early Middle Ages saw the beginning of the construction of the wall of dikes at the low-tide level, the pumping out of the water within the dikes, and the filling in of the low, tidal flats. The construction of new dikes beyond that point continued this process so that over the period of about 1,200 years, the Netherlands essentially doubled in size, greatly increasing the land available for farming. However, as reclaimed former sea bottom, much of this land (called polders by the Dutch) is below sea level. It is not uncommon to be in a Dutch field and

look up to the top of the dikes, several stories above. A breach in the dike wall would be a national disaster of tremendous magnitude. Vening Meinesz's gravity study was intended to be a temporary aside before an engineering career; however, the subject and its details fascinated the young man, and he spent the rest of his career in this discipline.

This career change became permanent when Vening Meinesz undertook a study of gravity and, in 1911, was part of a commission that mapped significant large areas; for example, the commission conducted a careful study of territory for a French Army base. By 1921, Vening Meinesz had created detailed gravity maps of the entire Netherlands landmass, using the technique he had developed for his doctoral work at the Technical University in Delft in 1915. He used two pendulums in surveying, whereas most surveyors used one. The two-pendulum method places the pendulums in perpendicular planes, canceling out variations in readings by taking averages.

By 1923, Vening Meinesz was ready to make gravity measurements of the Earth and to try his method at sea. The gravity measurements from the Earth's surface depend on the distance from the surface to the Earth's center. On land, that can vary considerably; an object on a high mountain weighs less than at sea level. Since the planet surface is largely water, and to deliver a complete picture, gravimetric mapping has to be carried out on all representative parts of the globe, measurements at sea were necessary. Vening Meinesz planned to use a submarine to reduce the effect of the waves on the pendulums, since that is inversely related to the depth below the surface. The first voyage in a submarine to Java in 1923 was not successful. Wave action still made the ship—even immersed in water and less buffeted by waves than one on the surface—a very unsteady platform for the pendulums, and the results were imprecise.

A 1925 expedition to Alexandria, Egypt, included a second attempt using three interconnected pendulums, and this method worked. Vening Meinesz achieved a successful gravity mapping of the geoid. The results showed that the Earth is not spherical: it is flattened at the poles, and bulges at the equator as a result of its rotation.

A university appointment in Delft in 1927 brought Vening Meinesz the chair in geodesy, cartography, and geophysics. He continued his travels and mapping; in all, there were 11 trips and 843 gravity determinations at different places at sea between 1923 and 1939. His most significant publication was the reports of his voyages, published as *Gravity Expeditions at Sea*, a five-volume work that appeared between 1932 and 1960. He served as president of the Association géodésique internationale (1933–55) and of the International Union of Geodesy and Geophysics (1948–51). Significantly, he discovered while mapping gravity that it showed anomalous changes, in long arcs that coincided with deep-ocean trenches such as those found on the ocean side of Japan and the West Indies. Since the sea is all at the same level, one might expect that the gravity measurements anywhere on the sea would be constant. They are not. These anomalies seemed to be related to the convection currents theorized by ARTHUR HOLMES in 1929. Holmes had proposed that these deep oceanic trenches were carved into the rock of the sea bottom by powerful, sediment-carrying scouring currents.

Vening Meinesz retired to Amersfoort around 1963, shortly before the University of Utrecht opened an Institute for Geophysics and Geochemistry named in his honor. He died in Amersfoort on August 11, 1966.

⊠ **Verrill, Addison Emery**
(1839–1926)
American
Taxonomist, Zoologist

Addison Verrill was born in Greenwood, Maine, on February 9, 1839, to George Washington and Lucy Hilborn Verrill. An early interest in collecting brought him to the Lawrence Scientific School (Harvard's science institute before science entered the Harvard curriculum), where LOUIS AGASSIZ was the principal instructor. A promising naturalist, Verrill began to work as one of Agassiz's assistants in 1860 while he was still a student. He graduated in 1862 and continued as an assistant in the Museum of Comparative Zoology in Cambridge, Massachusetts. When Yale University created a zoology department in 1864, Agassiz was asked to suggest a chairman, and he chose Verrill. Beginning as professor and chairman, Verrill remained in that position until his retirement in 1907.

In the fashion of his day, Verrill held other posts while serving as the zoology chair and principal researcher at Yale. He was curator of the Peabody Museum, Yale's natural history collection, from 1867 to 1910, and curator of the Boston Society of Natural History's collection, as well as the scientific chief of the U.S. Commission of Fish and Fisheries from 1871 to 1887. For three years (1868–70) Verrill taught comparative anatomy during the spring term at the University of Wisconsin. He was the associate editor of the *American Journal of Science* for 50 years (1869–1920), a member of the National Academy of Science, and the Société Zoologique de France, and president of the Connecticut Academy of Arts and Sciences.

Addison Verrill began his research in 1873 with SIDNEY SMITH, his brother-in-law. They examined and classified crustacea and produced three monographs, which Verrill finished after Smith became blind. In all, Verrill classified over 1,000 multicellular marine invertebrates. While working with KATHERINE BUSH, he concentrated on mollusks and cephalopods (squid, octopi, and sepia). Bush was a researcher who started as a "sorter girl" working for the Fish Commission. She became Verrill's assistant before enrolling as a student in zoology. The organisms studied ranged from those in the Vineyard Sound, in Massachusetts, to those in the Bay of Fundy, on the east coast of Canada, and south to Long Island Sound and eventually to the waters off Bermuda. Two of Verrill's best-known works were *The Bermuda Islands*, a two-volume work that appeared in installments between 1901 and 1907; and "A Report on Invertebrate Animals of Vineyard Sound and Adjacent Waters . . ." in the *Report of the U.S. Commission of Fish and Fisheries*.

His prodigious output of more than 300 papers was published almost entirely by either the *Report of the U.S. Commission of Fish and Fisheries* or the *Transactions of the Connecticut Academy of Arts and Sciences*. The subjects ranged from works on starfish of the Pacific coast in 1914 and those of Florida, the West Indies, and Brazil in 1915, to coelenterates of the Canadian coast, Bermuda, North and Central America, and Hawaii. After his retirement in 1907, Verrill went to the West Coast, where he continued his marine interests until he died in Santa Barbara, California, on December 10, 1926.

⊠ Vine, Frederick John

(1939–   )
British
Geophysicist

Frederick Vine, son of Frederick Royston and Ivy (Bryant) Vine, was born in London on June 17, 1939. He attended the Latymer School (Ham-

mersmith, London). His university training was all at Cambridge; he received a B.S. in physics from St. John's College, and his Ph.D. was granted in 1965. Upon graduation, Vine was appointed instructor in geology. A short stay in the United States from 1967 to 1970 was spent at Princeton University in New Jersey, where Vine was an assistant professor. He returned to Cambridge with the rank of reader (associate professor) and rose to a deanship (1977–80). Associated with the University of East Anglia from 1974 until his retirement in 1998, Vine was professor in the school of environmental sciences and recipient of the Balzan Prize in 1981.

The work that made Vine famous came very early in his career; he was still a graduate student and working in 1962 with DRUMMOND MATTHEWS. The paper detailing their findings was "Magnetic Anomalies over Oceanic Ridges," published in *Nature* in 1963. This work tied together many pieces of evidence that had been produced by researchers as varied as FELIX VENING MEINESZ, ALFRED WEGENER, ARTHUR HOLMES, ROBERT DIETZ, HARRY HESS, and W. MAURICE EWING, among others. If the seafloor is the youngest rock and spreads from the mid-ocean ridges, Vine and Matthews suggested, then each time new molten rock emerges from the ridge it pushes aside the already cooled rock that preceded it. Molten rock, coming up from the iron-rich interior of Earth, contains iron in the form of magnetite, and as it cools or crystallizes, the crystals align themselves with the Earth's magnetic field. If the Earth's magnetic field does from time to time reverse its polarity—the North Magnetic Pole becomes the South Magnetic Pole and vice versa, then the seafloor rocks should exhibit an alternating pattern of crystals, rather like zebra stripes. They do. Further study by Matthews, Vine, J. TUZO WILSON, and others showed that the time period for these reversals in polarity is about every half million years. Further work in 1966 used the magnetic strata on land to predict what the stripes on the seafloor should look like, and it was successful, confirming at the same time the theory of seafloor spreading and Wegener's idea of continental drift. This work was the capstone in what has been called a "revolution in science"—a totally new way to look at a phenomenon.

Frederick Vine, in addition to transmitting his enthusiasm for his subject to another generation of students, continued to explore the effect of the spread of the seafloor and how it changed the face of the Earth. He used the geomagnetic studies to work backward and created maps of the world in earlier geologic times. Examinations of the outline of Pangaea (the primordial landmass on Earth) and the velocity of movement of basement basalts produced a range of papers. Some of this work was detailed in "Spreading of Ocean Floor—New Evidence," published in *Science* in 1966; "Use of Paleomagnetism in Elucidating History of Ocean Basins," (1972) and "Implications of Proposed Geomagnetic Reversal Time-Scales for Early Tertiary and Late Cretaceous Periods" (1976), in *EOS Transactions of the American Geophysical Union;* and "Pangaea and the Geomagnetic Field in Permo-Triassic Times," published in *Geophysics: Journal of the Royal Astronomical Society* (1983).

Von Damm, Karen
(1955–   )
American
Geochemist

Karen Von Damm received all her university education in New England. She graduated from Yale University in 1977 with a B.S. in chemistry, and earned her Ph.D. in geochemistry in 1984, jointly from Massachusetts Institute of Technology (MIT) and Woods Hole Oceanographic Institution (WHOI), in Massachusetts. A postdoctoral research position took her to Menlo Park, California, for almost two years. There she was a National Research Council research associate in the U.S. Geological Survey's branch concerned with Pacific marine geology. This was

followed by a permanent post at Oak Ridge National Laboratory in Tennessee, where she was in the environmental sciences division. This work was concerned with contamination of river and estuary ecosystems. Since 1991, Von Damm has been professor of geochemistry and earth, oceans, and space (EOS) at the University of New Hampshire.

In her own description, "I am someone who is at the boundary of chemistry and geology/geophysics and this background has worked well for me in my studies of seafloor hydrothermal systems." Von Damm studies the chemistry of hot springs. She found a spring in 1991 at 9° North Latitude in the Pacific, at "time zero," the beginning of the eruption of that volcano. Since then, Von Damm and her students have visited the site repeatedly to observe its chemical changes as it matures. The biota that live on the site also change; the organisms that utilize the hydrogen sulfide–rich water emerging from a "black smoker" (an active vent) differ from the organisms that inhabit more diffuse sulfide-containing water farther away. Von Damm is also involved in the sampling of waters near the first of the black smokers to have been discovered in the tropical Pacific Ocean in the mid 1970s. Although water-rock interactions affected by heat and chemical composition occur more quickly in the black smokers, they take place in all waters. Von Damm is concerned with these water-rock reactions.

In 1993, Von Damm was part of a group using the *Alvin* (Woods Hole's submersible) to investigate the Lucky Strike hydrothermal site on the Mid-Atlantic Ridge. This is a slow-spreading site, unique because it is on a seamount. Her "The Geochemical Relationships Between Vent Fluids From the Lucky Strike Vent Field, Mid-Atlantic Ridge," published in *Earth & Planetary Sciences Letters* in 1998 arose from this work. Again using the *Alvin,* she explored the East Pacific Rise in 1998. This is a fast-spreading ridge with more than 40 new vents. Some, notably the Brandon, have super-hot water (405° C), and interesting chemical fluid mixtures. She published "Chemistry of Hydrothermal Vent Fluids from 9–10° N, East Pacific Rise: 'Time zero' the Immediate Post-eruptive Period," in the *Journal of Geophysical Research* in 2000.

In 2000, another group, including Von Damm, worked on the Gorda Ridge (off the coast of Oregon), sampling vent fluids and using the Monterey Aquarium's remote ocean vehicle (ROV) *Tiburon*. The expedition in 2001 used the *Jason* (the ROV belonging to WHOI) to explore hydrothermal sites in the Indian Ocean. Von Damm continues to study the vent communities in both Atlantic and Pacific Oceans. The chemistry at these sites is rapidly changing, and the communities of animals that depend on this mineral-rich water continue to provide many areas of scientific interest. Her papers on the geochemistry of vents and their attendant communities have appeared in noted journals such as *American Journal of Science*, *Earth & Planetary Sciences Letters*, and *Geophysical Research*.

W

Wallich, George Charles
(1815–1899)
British
Microbiologist

George Wallich, the son of Nathaniel Wallich, the superintendent of the Botanical Gardens in Calcutta, was born in Beverley in the north of England, in November of 1815. He attended the Reading Grammar School, and then went to Aberdeen, where he was most interested in art classes. Eventually, however, he realized that he could not earn his living as an artist. He enrolled in the medical faculty at the University of Edinburgh, where he was a classmate of EDWARD FORBES, whose work he later refuted. They received their degrees in 1836 and, unlike Forbes, Wallich worked as a doctor. He served as an army surgeon in India from 1838 until 1857, when he was disabled and sent back to England. After his recovery, he settled in Kensington, an area in London, and became a part of the London scientific discussion groups.

His last official assignment was in 1860, when he served as the naturalist aboard the survey ship HMS *Bulldog*. The charge was to find a route for the transatlantic telegraph cable. It also marked the beginning of Wallich's career as a marine scientist.

While aboard the *Bulldog*, he was in charge of the depth soundings that brought up tiny starfish from depths greater than 1,260 fathoms (2,291 m or 7,560 ft). According to Forbes and his followers, the ocean at this depth should definitely have been an azoic zone, a region with no life in it. Wallich became an early and vocal refuter of the "azoic zone" concept. His paper, "Existence of Animal Life at Great Depths in the Sea" (1860) in *Annals and Magazine of Natural History* was one of several that dealt with this subject.

Wallich achieved local fame when he won an argument; Thomas Huxley had maintained that the contents of a murky vial was a primitive organism. He named it for ERNST HAECKEL. Wallich, who had become an expert on the protozoa, demonstrated that it was not alive. Even Huxley had to agree. Later, John Young Buchanan, the chemist on the *Challenger* voyage that studied the world's oceans in 1872–76, found that the purported organism was calcium sulfate precipitated by alcohol that was added to the seawater to preserve the organism. Wallich continued to contribute works on protozoa and attend scientific meetings. "On Vital Functions of Deep Sea Protozoa," published in 1969 in the *Monthly Microscopial Journal*, and "Radiolarians as an Order of Protozoa," published in 1978 in *Popular Science Review*, attest to his continuing efforts. He also described the state of biology as people in that field reacted to the concept of evolution, in "Threshold of Evolution," published in 1880, also in the *Popular Science Review.* Wallich was honored by the Linnaean Society of London in 1898, when they named him recipient of the society's annual medal. He died on March 31, 1899 in his London home.

⊠ **Wegener, Alfred Lothar**
(1880–1930)
German
Meteorologist, Geophysicist

Alfred Wegener was born on November 1, 1930, in Berlin, Germany, to Richard and Anna Wegener. His father was the director of an orphanage. The boy entered the University of Berlin with a career as an astronomer in mind; he received his Ph.D. in astronomy in 1904. He had always expressed interest in geophysics, however, and when he started looking for a position, meteorology caught his attention. At the time, it was an expanding field. Wegener pioneered a method of tracking air currents and circulation with balloons, and wrote a text on meteorology that became the standard for the field. *Thermodynamik der Atmosphäre* (Thermodynamics of the atmosphere), published in 1911. His work on air circulation gained him attention, and he was invited to join an expedition to Greenland to study polar air currents. The expedition left in 1906 and returned in 1908. By then he had sufficient reputation to warrant an offer from the University of Marburg to become a tutor there. Four years later, another expedition to Greenland took him away for more than a year. By 1914, Wegener had been drafted into the German Army. He spent most of the war years in the army's weather service after having been wounded in combat.

Returning to Marburg after the war was frustrating for Wegener. There were very limited funds available for research, and he spent much time in bureaucratic maneuvering for the money. A post was especially created for him in meteorology and geophysics at the University of Graz (Austria), and the family moved there in 1924.

By the early 1900s, a number of geographers, geologists, and mariners had remarked on the complementary nature of the continental coasts: the eastcoast of South America meshes nicely with the west coast of Africa. Beginning about 1910, Wegener began to construct a theory explaining why this was so. He redeveloped EDUARD SUESS's ideas that all the continents were, in the late Paleozoic (about 250 million years ago), one single landmass, a super-continent Suess had referred to as Pangaea. By 1911, Wegener came upon a scientific paper that listed identical fossil plants and animals in South America and Africa. The geological explanation of the day for these finds was that biota had moved from one continent to another by way of land bridges, and that rising ocean water levels have since caused these to become submerged. Observing that the distance was too great and the ocean bottom deep rather than shallow, Wegener dismissed this explanation.

Wegener developed theories to explain the appearance of similar biota on widely separated landmasses; the similarities of the Appalachian Mountains and the Highlands in Scotland, the Karroo rock formations in South Africa and the identical ones in Santa Catarina (Brazil), and the fossils of tropical plants found near Spitsbergen, Norway. He theorized that the continents had moved slowly away from each other and that this process took millions of years. He first presented this material in support of what he called "continental displacement," now referred to as continental drift, in his 1912 lectures. The published version appeared in 1915 as *Die Entstehung der Kontinente und Ozeane* (The origin of continents and oceans). This was Wegener's most significant work; it was expanded and reissued in 1920, 1922, and 1929. His other significant work was a book on paleoclimatology, *Die*

Alfred Wegener *(Alfred Wegener Institute for Polar and Marine Research)*

Klimate der Geologischen Vorzeit (Climate of early geologic times), published with his father-in-law, Wladimir Koppen, in 1924.

The response to this work from the scientific community was almost uniformly negative. Leading figures in geology, such as Rollin Chamberlin at the University of Chicago and Harold Jeffreys of Cambridge University, were adamantly against this idea created by a scientific amateur—after all, Wegener was a meteorologist, not a geologist. In one very critical aspect, they were correct. Wegener's explanation of what propelled the crustal masses, the mechanism by which the continents drifted, was not persuasive, and he had them moving too fast. He theorized that the propelling force was tidal; he knew that strong ocean currents existed and assumed that their force would be sufficient. The weight of opposition and Wegener's early death meant that the theory all but disappeared. It might have become a scientific footnote if not for the efforts of ARTHUR HOLMES. But there were other supporters; Alexander du Toit in South Africa saw Wegener's work as an explanation for the related biota in Africa and South America, and the Swiss Emile Argand accepted the idea that the folded and buckled landscape of the Alps was the result of the upheaval from the collision of landmasses. It was not until the 1940s and 1950s that the work of ROBERT DIETZ, HARRY HESS, BRUCE HEEZEN, W. MAURICE EWING, J. TUZO WILSON, and others made a good case for continental drift. By then, the spreading seafloor was recognized as the moving force.

Wegener's last Greenland expedition began in the summer of 1930. A group of colleagues was stranded on the Greenland icecap with food running out. Wegener brought the replacement food cache and, on returning, was caught in a snowstorm and lost. He died some days after his 50th birthday.

Wilson, Edmund Beecher
(1856–1939)
American
Biologist

Edmund Wilson was born in Geneva, Illinois, on October 19, 1856, to Isaac and Caroline Wilson. He was brought up by an aunt and uncle, because his father, a circuit court judge, traveled most of the time, taking his wife with him. The boy grew up a collector. At 16, he substituted for his older brother as a schoolteacher, and the experience left him with several conclusions: he didn't want to repeat that year, he wanted to go to college, and he wanted to be a scientist. Helped by his father, Edmund sat for the West Point entrance examination, and passed it with a high score but was too young to be admitted. He spent the next year at Antioch College in Ohio, then another year at the University of Chicago, and in 1875 entered the Sheffield Scientific School at Yale. After receiving his bachelor's degree in 1878, he stayed on as a graduate student. His thesis for the master's degree was on sea spiders, *Pycnogonida*.

Wilson's cousin Samuel Clarke, a student at Johns Hopkins University in Baltimore, Maryland, persuaded Wilson to enroll. There he worked with WILLIAM BROOKES, a morphologist who was looking into development of body structures. This was the beginning of embryology, and Brookes passed the enthusiasm for research and new information to his students. In 1881, Wilson completed his work for a Ph.D. in zoology, and went abroad on a trip financed by his older brother. The object was to visit other contemporary research biologists. On his return, he did a year of substitute teaching again, this time for his cousin, Samuel Clarke, at Williams College in Massachusetts (1883–84). In the same year (1883), his doctoral dissertation and first major work, "The Development of *Renilla*" (a colonial polyp), was published in *Philosophical Transactions of the Royal Society of London*. In this long paper, he began his lifelong study of the development of embryonic cells.

After voicing their complaints about the lack of time and material with which to do research at Williams College, Wilson and a friend (W. T. Sedgwick) began working on a book that was finished the next year, when Wilson was teaching at the Massachusetts Institute of Technology (1884–85). *General Biology* (1886) became the standard textbook in the field. Wilson's next move was to teach at the new women's college in Bryn Mawr, Pennsylvania, where he was chair of the biology department (1885–91). This department was also known for the work done there by JACQUES LOEB and DAVID TENNENT, both researchers in embryology. When Wilson moved again, in 1891, it was to Columbia University, where he was professor of zoology until his retirement in 1928.

Building on work done on earthworms (terrestrial annelids) in the late 1880s, Wilson produced several papers on *Nereis*, a marine annelid (1892, 1895, 1898), in which he followed the development of "cell lineage." In other research, Wilson divided developing embryos into two large groups after they had developed three germ layers (the gastrula). As the fertilized egg's cells divide, they first form a hollow ball, a blastula; then the ball develops a cleft, and three germ layers of cells form from the in-folding. These three layers start the differentiation of the future organism's systems. One of Wilson's groups were those embryos that produced a mesoderm (middle layer) in a mosaic or spiral fashion, or those in which the mesoderm developed radially, growing outward from the center. Annelids (segmented worms), arthropods (shrimp, lobsters, and insects), and mollusks (snails, squid, oysters, and clams) all belong to the mosaic path of development, and echinoderms (starfish), and chordates, including vertebrates, develop radially. From this platform in descriptive embryology, Wilson's research moved to a more experimental approach in "Studies on Chromosomes I . . .," published in the *Journal of Experimental Zoology* in 1905. This paper and those that followed it are on the discovery and development of sex chromosomes.

Taking his start from ideas that were current when he toured European universities, Wilson's experimental program led him to explore the controversy of cell development: Did it proceed by epigenesis or preformation? In epigenesis a germ cell is undifferentiated; in preformation, all the organs and structures that the organism will have are present in the one-cell stage. Wilson leaned toward the latter explanation. This work was collected and published as *The Cell in Development and Inheritance* (1896), an important book that was revised and augmented through several editions. Wilson was an early adherent of Gregor Mendel, the experimenter who first explained the laws of inheritance based on his many years of tracking the characteristics of garden peas. His work emphasized the importance of chromosomes in heredity, but the last revision of his book in 1925 stressed the importance of other cellular structures that determine heredity, notably mitochondria. Mitochondria are structures within cells that are the metabolizing agents; they make energy available in the cell. They also carry nonnuclear genetic material, which is inheritable.

Wilson spent his last years in New York City, where he died on March 3, 1939. The department he built was a base for the brilliant work of Hermann Muller and Thomas Hunt Morgan and others, who developed the study of inheritance that led to contemporary work on genome research. Muller and Morgan were the geneticists who proved that Mendel's principles of inheritance were applicable to all organisms.

Wilson, John Tuzo
(1908–1993)
Canadian
Geophysicist

Tuzo Wilson (as he was commonly called) was born on October 24, 1908, in Ottawa, Canada, to John and Henrietta Tuzo Wilson. His engineer father had emigrated from Scotland. Tuzo Wilson studied geophysics and received an undergraduate degree in that subject in 1930 from the University of Toronto, where he was the first recipient of that degree at that university. After a two-year research period at Cambridge University, England, he began study in the Ph.D. program in geology at Princeton University in New Jersey, finishing in 1936. While at Princeton, Wilson was influenced by the brilliant young lecturer, HARRY HESS. Wilson spent the next three years working at various geological ventures. Army service began for Wilson in 1939 and continued throughout World War II; he was a colonel by 1946. After leaving the army he joined the faculty at the University of Toronto, staying there as professor of geophysics until he retired in 1974.

The work of Tuzo Wilson in advancing the theory of plate tectonics was crucial. ALFRED WEGENER's concept of mobile fragments of crust floating on the upper layer of Earth, some bearing continents, others oceans, combined with Harry Hess's work on seafloor spreading, allowed Tuzo Wilson to produce a variety of papers supporting these ideas. By the 1960s, he had developed his own ideas concerning island chain formation. He hypothesized that islands, such as the Hawaiian Islands and the Japanese island chain, were formed as a plate moved over a "hot spot" that was stationary in the underlying mantle layer. This theory explained the occurrence of volcanic islands in places far from the nearest plate edge, where most other activity was obvious. His paper proposing this idea, "A Possible Origin of Hawaiian Islands," was rejected by all the major international journals and finally published by the *Canadian Journal of Physics* in 1963.

This revolutionary work was followed two years later by "A New Class of Faults and Their Bearing on Continental Drift," published in *Nature*; then major journals were no longer dismissing his work. With FREDERICK VINE as coauthor, he wrote "Magnetic Anomalies Over a Very Young Ocean Ridge off Vancouver Island," published in *Science* in 1965. These were followed by "Is the African Plate Stationary?" in *Nature* in 1972, and "Continental Drift Emphasizing the History of the South-Atlantic Area," in *Transactions of the American Geophysical Union* in the same year.

Wilson proposed a third type of plate boundary to add to his first two that were accepted, spread or plate growth as new material is accreted, and the phenomenon of one plate overriding another, pushing it down into the mantle. The third type of plate boundary connects ocean ridges and trenches, which can end abruptly and transform themselves into major faults that slide horizontally. The San Andreas fault in California is such a transform fault boundary. The Pacific plate is moving northward against the North

American plate. In popular terms, western California is moving to Alaska. Such faults move crust laterally, without creation of new material or destruction of existing crust.

Wilson continued to publish papers building on these concepts and others, notably the dating of island arcs by their distance from the plate edge, well into his "retirement." He then moved his research interests to environmental problems and produced a body of work in that area, serving as director of the Ontario Science Centre. Other administrative posts included the presidency of the Royal Society of Canada from 1972 to 1973, and of the American Geophysical Union from 1980 to 1982. Wilson then became chancellor of York University (Toronto) from 1983 to 1986. After Wilson's death in Los Gatos, California, on April 15, 1993, a mountain range in Antarctica was named in his honor.

⊠ **Wüst, Georg Adolf Otto**
(1890–1977)
German
Oceanographer

Georg Wüst was born on June 15, 1890, in Posen, Germany (now Poznan, Poland). The prestigious University of Berlin was his destination for graduate education, and he finished his doctorate in physics there in 1919. At that point, Germany was recovering from World War I and the imposition of a heavy war debt. In response, a group of German scientists developed a project to explore the possibility, suggested by the Nobel Prize–winning chemist, Fritz Haber, that gold could be profitably extracted from seawater and used as a source of revenue. Using the research and survey vessel *Meteor* for the expedition, the group had to acknowledge that finding and refining what turned out to be minute amounts of gold was not a financially viable plan. They dismissed that part of their mission.

The director of the project, Alfred Merz, was also director of the Institut für Meerskunde (Oceanographic Institute) in Kiel, Germany. He had been Wüst's mentor at the University of Berlin, greatly extending the scope of this expedition. When he died unexpectedly, Wüst became chief oceanographer on the voyage, which undertook an extensive exploration of the Atlantic Ocean (1925–27). The *Meteor* expedition probably marked the beginning of physical oceanography observations on a global scale. It was the first attempt to look at an ocean as an entity, studying every aspect of it from meteorology to water circulation, chemical analysis, biotic surveys, and bottom sampling. The *Meteor* expedition is especially important because carefully thought out observing plans were combined with meticulous observations, thorough analysis and publication. Each part of the study was connected to the others; the chemistry studies were directly affected by and related to the biological studies, and the physics research was combined with the chemistry work. Wüst, who had unexpectedly assumed a leadership role during the *Meteor* expedition, must be recognized for the meticulous standards he set for the analysis of the data. This work was published as *The Atlas of the Stratification and Circulation of the Atlantic Ocean* by the National Science Foundation of the United States in 1993. It had achieved international fame long before its final publication.

Wüst led a second voyage to the Atlantic in 1938. The second *Meteor* expedition was an international undertaking to explore the Gulf Stream. While the "river in the ocean" had been described earlier by many, BENJAMIN FRANKLIN was the first to name it in a written work. This trip, probably the last international cooperative effort before the beginning of World War II, was a systematic scientific study.

After the war, the Oceanographic Institute in Kiel was in ruins. Wüst spent his energies on the reestablishment of this institution as a scientific research center. He worked as its director from 1946 until his retirement in 1959, when he moved to Erlangen, a university city in southern Germany. Wüst died there on November 8, 1977.

His work on the *Meteor* expeditions earned him many accolades, all deserved. The later effort to rebuild the institute in Kiel was a labor of love, both for his teacher and his subject.

Z

Zubov, Nikolay Nikolaevitch
(1885–1960)
Russian
Oceanographer

The family of Nikolay Zubov, who was born in Izmail (now in the Ukraine) on May 23, 1885, was one that had several members in the Czarist navy. It was assumed that Nikolay would follow suit, and in 1904 he was accepted by the Czar's Naval Cadet Corps School. He continued in navy schools and graduated from the Russian Naval Academy in 1910 with a degree in hydrography.

Scientific exploration was an early part of Zubov's career. In 1912, he was on the *Bakan* survey of an Arctic Ocean inlet on Novaya Zemlya. The next year, he worked for the civil service as the hydrographer on another Arctic survey, and in 1914, the government sent him to the Geophysical Institute in Bergen, Norway, the European leaders in meteorology.

Naval service was the major component of Zubov's life between 1914 and 1918, but after World War I, he was an early member of the Floating Marine Science Institute. That body was an informal group for several years before it was formally established in 1921. Its name was later changed to the State Institute of Oceanography, and its charge was almost exclusively the study of the Arctic.

While on extended leave from the navy, Zubov made four trips to the Barents Sea. The first, in 1923, was on board the *Persia*. He published a number of scientific papers on his findings there and, finally, a memoir in 1932. The papers described his lifelong interest in vertical mixing of ocean water, specific currents, and the nature and construction of sea ice. It is different in various locations, and an important tool in meteorology and long-term weather studies. In his papers, he took the works of VILHELM BJERKNES well beyond what Bjerknes had called "convection currents." Zubov's handbook on this subject, *Oceanographic Tables*, published in 1935, resulted in Russian mapping of the Arctic waters and their currents.

The name Zubov gave his field of study was oceanology, not oceanography. He believed that his work was quantitative and that oceanography was merely descriptive. In spite of his efforts to make a distinction, the field has remained oceanography to everyone else, although to some minds, it is physical oceanography. However, the department Zubov headed at the Hydrometeorological Center in Moscow, created in 1931, was the department of oceanology.

Zubov was secretary of the Soviet committee of the Second International Polar Year (1932–33). His oceangoing voyage was part of that year's program; he was on the NM *Knipovich*, the first ship to sail around Franz Josef Land. Zubov investigated sea ice and the effects of climate and currents on its depth in a 1935 exploration of northern Greenland and the Barents and Kara Seas, while on board the *Sadko,* an icebreaker. There, Zubov charted the warm currents of the Atlantic Ocean and their effects on the waters of the Kara Sea. He also studied the use of aerial reconnaissance in navigation. In World War II, Zubov returned to the navy and was assigned to the Arctic. This led to an important book, *Arctic Ice,* on the dynamics of ice and its relationship to the surrounding air and water, published in 1945.

After World War II, Zubov was professor of oceanology at the University of Moscow from 1949 until his death on November 11, 1960. The last scientific paper he delivered was on the relationship of barometric pressure and sea level. It was presented in 1959 at the first International Oceanographic Congress in New York City and was well received. Given the hostile relationship between the Soviet Union and the west, his appearance was remarkable. Zubov published over 200 works; some were translated into English. They include *An Elementary Study of High Tides,* 1933; *Seawater and Sea Ice,* 1938; *In the Middle of the Arctic,* 1948; and *A Basic Study of Straits of the World's Oceans,* 1956.

Entries by Field

ARCHEOLOGY
Bass, George

ASTRONOMY
Bowditch, Nathaniel
Halley, Edmond

BIOLOGY
Audouin, Jean-Victor
Brookes, William Keith
Buffon, Georges-Louis LeClerc,
    comte de
Bush, Katherine
Carson, Rachel Louise
Conklin, Edwin Grant
Darwin, Charles Robert
Earle, Sylvia Alice
Ehrenberg, Christian Gottfried
Haeckel, Ernst Heinrich
    Philipp August
Kofoid, Charles Atwood
Loeb, Jacques
Lovén, Sven Ludwig
Morse, Edward Sylvester
Murray, Sir John
Pourtalès, Louis-François de
Quatrefages de Bréau, Jean-
    Louis-Armand de
Rathbun, Mary Jane
Rondelet, Guillaume

Schafer, Wilhelm
Schmidt, Ernst Johannes
Tennent, David Hilt
Verrill, Addison Emery
Wallich, George Charles

BOTANY
Cleve, Per Teodor
Dujardin, Félix
Kylin, Johann Harald
Marion, Antoine-Fortuné
Murray, George Robert Milne

CARTOGRAPHY, EXPLORATION
Bellingsgauzen, Faddei
Bougainville, Louis-Antoine,
    comte de
Bowditch, Nathaniel
Charcot, Jean-Baptiste
Cook, James
Dumont d'Urville, Jules-
    Sébastien-César
Forster, Johann Reinhold
Humboldt, Friedrich Wilhelm
    Karl Heinrich Alexander,
    baron von
Maury, Matthew Fontaine
Nansen, Fridtjof
Penck, Albrecht
Ross, Sir James Clark

Scoresby, William
Tharp, Marie

CHEMISTRY
Cleve, Per Teodor
Dittmer, William
Forchhammer, Johann Georg
Scranton, Mary Isabelle

COMPARATIVE ANATOMY
Cuvier, Georges, Baron
Geoffroy Saint-Hilaire, Étienne
Goodrich, Edwin Stephen
Müller, Johannes Peter
Sabatier, Armand

EMBRYOLOGY
Conklin, Edwin Grant
Delage, Yves
Ehrenberg, Christian Gottfried
Loeb, Jacques

ENVIRONMENTAL SCIENCE
Brookes, William Keith
Cousteau, Jacques
Earle, Sylvia Alice
McCammon, Helen Mary
Mann, Kenneth H.

GEOLOGY
Bailey, Sir Edward Battersby
Dana, James Dwight
Darwin, Charles Robert
Desor, Pierre
Forchhammer, Johann
    Georg
Godwin-Austen, Robert
    Alfred Cloyne
Heezen, Bruce Charles
Helland-Hansen, Bjørn
Hess, Harry Hammond
Holmes, Arthur
Johnson, Douglas Wilson
Knipovitch, Nikolai
    Mikhailovitch
Kofoid, Charles Atwood
Kuenen, Philip Henry
Lyell, Sir Charles
Maillet, Benoît de
Marion, Antoine Fortuné
Matsuyama (Matuyama),
    Motonori
Menard, Henry William
Mesyatsev, Ivan Illarionivich
Mohorovičić, Andrija
Moro, Antonio Lazzaro
Neumayr, Melchior
Nordenskjöld, Nils Adolf Erik
Renard, Alphonse
Suess, Eduard
Sverdrup, Harald Ulrik
Vine, Frederick John
Wegener, Alfred Lothar

GEOPHYSICS
Bowie, William
Duperrey, Louis-Isidore
Ewing, William Maurice
Ferrel, William
Halley, Edmond
Hunt, Maurice Neville

Matsuyama (Matuyama),
    Motonori
Matthews, Drummond Hoyle
Vening Meinesz, Felix Andries
Wilson, John Tuzo

HISTORY
Cuvier, Georges, Baron
Nordenskjöld, Nils Adolf Erik

HYDROGRAPHY, HYDROLOGY
Duperrey, Louis-Isidore
Knipovitch, Nikolai
    Mikhailovitch

MARINE BIOLOGY
Agassiz, Alexander
Knipovitch, Nikolai
    Mikhailovitch
Lang, Arnold
Rathbun, Mary Jane
Sars, Michael

MARINE GEOLOGY
Andrusov, Nikolai Ivanovich
Ballard, Robert Duane
Dietz, Robert Sinclair
Ewing, William Maurice
Heezen, Bruce Charles
Hess, Harry Hammond

MATHEMATICS
Bernoulli, Daniel
Bowditch, Nathaniel
Cardano, Girolamo
Ekman, Vagn Walfrid
Ferrel, William
Shtokman, Vladimir
    Borisovich

METEOROLOGY
Albert I
Bjerknes, Jakob Aall Bonnevie

Bjerknes, Vilhelm Frimann
    Køren
Ferrel, William
Franklin, Benjamin
Maury, Matthew Fontaine
Sverdrup, Harald Ulrik
Zubov, Nikolay Nikolaevitch

NATURAL HISTORY, NATURAL SCIENCE
Agassiz, Louis
Anning, Mary
Aristotle
Bede
Beebe, Charles William
Bory de St. Vincent, Jean-
    Baptiste-Geneviève-
    Marcellin
Bougainville, Louis-Antoine,
    comte de
Buffon, George-Louis LeClerc,
    comte de
Cardano, Girolamo
Carpenter, William Benjamin
Dumont d'Urville, Jules-
    Sébastien-César
Forster, (Johann) Georg Adam
Forster, Johann Reinhold
Goethe, Johann Wolfgang von
Humboldt, Friedrich Wilhelm
    Karl Heinrich Alexander
    baron von
Lacépède, Bernard-Germain-
    Étienne de la Ville-sur-
    Illion, comte de
Lamarck, Jean-Baptiste-Pierre-
    Antoine de Monet,
    chevalier de
Lesueur, Charles-Alexandre
Marsigli, Count Luigi
    Ferdinando
Rondelet, Guillaume

OCEANOGRAPHY
Agassiz, Alexander
Albert I
Bellingsgauzen, Faddei
Derugin, Konstantin
    Mikhailovich
Ekman, Vagn Walfrid
Feely, Richard Alan
Hensen, Viktor
Maillet, Benoît
Makarov, Stepan Osipovich
Menard, Henry William
Mesyatsev, Ivan Illarionivich
Munk, Walter Heinrich
Pettersson, Hans
Revelle, Roger Randall
    Dougan
Smith, Deborah K.
Thomson, Sir Charles Wyville
Wüst, Georg Adolf Otto
Zubov, Nikolay Nikolaevitch

PALEONTOLOGY
Agassiz, Louis
Anning, Mary

Cushman, Joseph
Cuvier, Georges, Baron
Desor, Pierre
Moro, Antonio Lazzaro
Neumayr, Melchior
Schafer, Wilhelm

PARASITOLOGY
Leuckart, Karl Georg Friedrich
    Rudolf
Wallich, George Charles

PHYSICS
Bache, Alexander Dallas
Franklin, Benjamin
Hunt, Maurice Neville
Lenz, Heinrich Friedrich
    Emil
Medwin, Herman
Munk, Walter Heinrich
Piccard, Auguste
Poli, Giuseppe Saverio
Vening Meinesz, Felix
    Andries
Wüst, Georg Adolf Otto

PHYSIOLOGY
Audouin, Jean-Viktor
Haldane, John Scott
Hensen, Viktor
Portier, Paul
Richet, Charles Robert

ZOOLOGY
Buffon, Georges-Louis LeClerc,
    comte de
Bush, Katherine
Cuvier, Georges, Baron
Delage, Yves
Forbes, Edward
Jeffreys, John Gwyn
Lang, Arnold
Lankester, Sir Edwin Ray
Leuckart, Karl Georg Friedrich
    Rudolf
Milne-Edwards, Henri
Möbius, Karl August
Rathbun, Mary Jane
Smith, Sidney Irving
Thompson, John Vaughan
Verrill, Addison Emery

AUSTRIA
Munk, Walter Heinrich

BELGIUM
Renard, Alphonse

CANADA
McCammon, Helen Mary
Murray, Sir John
Wilson, John Tuzo

CROATIA
Mohorovičić, Andrija

DENMARK
Forchhammer, Johann Georg
Schmidt, Ernst Johannes

FRANCE
Albert I
Audouin, Jean-Victor
Bory de St. Vincent, Jean-
    Baptiste-Geneviève-
    Marcellin
Bougainville, Louis-Antoine,
    comte de
Buffon, Georges-Louis LeClerc,
    comte de

Chappe d'Auteroche, Jean-
    Baptiste
Charcot, Jean-Baptiste
Cousteau, Jacques
Cuvier, Georges Léopold-
    Chrétien-Frédéric-
    Dagobert, Baron
Delage, Yves
Dujardin, Félix
Dumont d'Urville, Jules-
    Sébastien-César
Duperrey, Louis-Isidore
Geoffroy Saint-Hilaire,
    Étienne
Lacépède, Bernard-Germain
    Étienne, de la Ville-sur-
    Illion, comte de
Lamarck, Jean-Baptiste-Pierre-
    Antoine de Monet,
    chevalier de
Lesueur, Charles-Alexandre
Maillet, Benoît de
Marion, Antoine-Fortuné
Milne-Edwards, Henri
Portier, Paul
Quatrefages de Bréau, Jean-
    Louis-Armand de
Richet, Charles Robert
Sabatier, Armand

GERMANY
Desor, Pierre
Dittmer, William
Ehrenberg, Christian Gottfried
Forster, (Johann) Georg Adam
Forster, Johann Reinhold
Goethe, Johann Wolfgang von
Haeckel, Ernst Heinrich
    Philipp August
Hensen, Viktor
Humboldt, Friedrich Wilhelm
    Karl Heinrich Alexander,
    baron von
Leuckart, Karl Georg Friedrich
    Rudolf
Loeb, Jacques
Möbius, Karl August
Müller, Johannes Peter
Neumayr, Melchior
Penck, Albrecht
Schafer, Wilhelm
Wegener, Alfred Lothar
Wüst, Georg Adolf Otto

GREAT BRITAIN
Anning, Mary
Bailey, Sir Edward Battersby
Bede
Carpenter, William Benjamin

Cook, James
Darwin, Charles Robert
Forbes, Edward
Godwin-Austen, Robert Alfred
    Cloyne
Goodrich, Edwin Stephen
Haldane, John Scott
Halley, Edmond
Holmes, Arthur
Hunt, Maurice Neville
Jeffreys, John Gwyn
Kuenen, Philip Henry
Lankester, Sir Edwin Ray
Lyell, Sir Charles
Mann, Kenneth H.
Matthews, Drummond Hoyle
Murray, George Robert Milne
Ross, Sir James Clark
Suess, Eduard
Thompson, John Vaughan
Thomson, Sir Charles Wyville
Vine, Frederick John
Wallich, George Charles

GREECE
Aristotle

ITALY
Cardano, Girolamo
Marsigli, Count Luigi
    Ferdinando
Moro, Antonio Lazzaro
Poli, Giuseppe Saverio

JAPAN
Matsuyama (Matuyama),
    Motonori

NETHERLANDS
Bernoulli, Daniel

Vening Meinesz, Felix
    Andries

NORWAY
Bjerknes, Jakob Aall Bonnevie
Bjerknes, Vilhelm Frimann
    Køren
Helland-Hansen, Bjørn
Nansen, Fridtjof
Sars, Michael
Sverdrup, Harald Ulrik

RUSSIA
Andrusov, Nikolai Ivanovich
Bellingsgauzen, Faddei
Derugin, Konstantin
    Mikhailovich
Knipovitch, Nikolai
    Mikhailovitch
Lenz, Heinrich Friedrich
    Emil
Makarov, Stepan Osipovich
Mesyatsev, Ivan Illarionivich
Shtokman, Vladimir
    Boresovich

SWEDEN
Cleve, Per Teodor
Ekman, Vagn Walfrid
Kylin, Johann Harald
Lovén, Sven Ludwig
Nordenskjöld, Nils Adolf
    Erik
Pettersson, Hans

SWITZERLAND
Agassiz, Alexander
Agassiz, Louis
Lang, Arnold
Piccard, Auguste

Pourtalès, Louis-François de

UNITED STATES
Bache, Alexander Dallas
Ballard, Robert Duane
Bass, George
Beebe, Charles William
Bowditch, Nathaniel
Bowie, William
Brookes, William Keith
Bush, Katherine
Carson, Rachel Louise
Conklin, Edwin Grant
Cushman, Joseph
Dana, James Dwight
Dietz, Robert Sinclair
Earle, Sylvia Alice
Ewing, William Maurice
Feely, Richard Alan
Franklin, Benjamin
Heezen, Bruce Charles
Hess, Harry Hammond
Johnson, Douglas Wilson
Kofoid, Charles Atwood
Maury, Matthew Fontaine
Medwin, Herman
Menard, Henry William
Morse, Edward Sylvester
Rathbun, Mary Jane
Revelle, Roger Randall
    Dougan
Scranton, Mary Isabelle
Smith, Deborah K.
Smith, Sidney Irving
Tennent, David Hilt
Tharp, Marie
Verrill, Addison Emery
Von Damm, Karen L.

ENTRIES BY COUNTRY
OF MAJOR SCIENTIFIC ACTIVITY

AUSTRIA
Neumayr, Melchior
Suess, Eduard

BELGIUM
Renard, Alphonse

CANADA
Mann, Kenneth H.
Wilson, John Tuzo

CROATIA
Mohorovičić, Andrija

DENMARK
Forchhammer, Johann Georg
Schmidt, Ernst Johannes

FRANCE
Audouin, Jean-Victor
Bory de St. Vincent, Jean-
    Baptiste-Geneviève-
    Marcellin
Bougainville, Louis-Antoine,
    comte de
Buffon, George Louis Leclerc,
    comte de
Chappe d'Auteroche, Jean-
    Baptiste
Charcot, Jean-Baptiste
Cousteau, Jacques

Cuvier, Georges, Baron
Delage, Yves
Dujardin, Félix
Dumont d'Urville, Jules-
    Sébastien-César
Duperrey, Louis-Isidore
Geoffroy Saint-Hilaire,
    Étienne
Lacépède, Bernard-Germain-
    Étienne de la Ville-sur-
    Illion, comte de
Lamarck, Jean-Baptiste-Pierre-
    Antoine de Monet,
    chevalier de
Lesueur, Charles-Alexandre
Maillet, Benoît de
Marion, Antoine-Fortuné
Milne-Edwards, Henri
Portier, Paul
Quatrefages de Bréau, Jean-
    Louis-Armand de
Richet, Charles Robert
Rondelet, Guillaume
Sabatier, Armand

GERMANY
Ehrenberg, Christian
    Gottfried
Goethe, Johann Wolfgang von
Haeckel, Ernst Heinrich
    Philipp August

Hensen, Viktor
Humboldt, Friedrich Wilhelm
    Karl Heinrich Alexander,
    baron von
Leuckart, Karl Georg Friedrich
    Rudolf
Möbius, Karl August
Müller, Johannes Peter
Penck, Albrecht
Schafer, Wilhelm
Wegener, Alfred Lothar
Wüst, Georg Adolf Otto

GREECE
Aristotle

GREAT BRITAIN
Anning, Mary
Bailey, Sir Edward Battersby
Bede
Carpenter, William Benjamin
Cook, James
Darwin, Charles Robert
Dittmer, William
Forbes, Edward
Forster, (Johann) Georg Adam
Forster, Johann Reinhold
Godwin-Austen, Robert Alfred
    Cloyne
Goodrich, Edwin Stephen
Haldane, John Scott

Holmes, Arthur
Hunt, Maurice Neville
Jeffreys, John Gwyn
Lankester, Sir Edwin Ray
Lyell, Sir Charles
Matthews, Drummond Hoyle
Murray, George Robert Milne
Murray, Sir John
Ross, Sir James Clark
Scoresby, William
Thompson, John Vaughan
Thomson, Sir Charles Wyville
Vine, Frederick John
Wallich, George Charles

ITALY
Cardano, Girolamo
Marsigli, Count Luigi
    Ferdinando
Moro, Antonio Lazzaro
Poli, Giuseppe Saverio

JAPAN
Matsuyama (Matuyama),
    Motonori

MONACO
Albert I

NETHERLANDS
Kuenen, Philip Henry
Vening Meinesz, Felix Andries

NORWAY
Bjerknes, Jakob Aall Bonnevie
Bjerknes, Vilhelm Frimann
    Køren
Helland-Hansen, Bjørn

Nansen, Fridtjof
Sars, Michael
Sverdrup, Harald Ulrik

RUSSIA
Andrusov, Nikolai Ivanovich
Bellingsgauzen, Faddei
Derugin, Konstantin
    Mikhailovich
Knipovitch, Nikolai
    Mikhailovitch
Lenz, Heinrich Friedrich Emil
Makarov, Stepan Osipovich
Mesyatsev, Ivan Illarionivich
Shtokman, Vladimir
    Boresovich

SWEDEN
Cleve, Per Teodor
Ekman, Vagn Walfrid
Kylin, Johann Harald
Lovén, Sven Ludwig
Nordenskjöld, Nils Adolf
    Erik
Pettersson, Hans

SWITZERLAND
Bernoulli, Daniel
Desor, Pierre
Lang, Arnold
Piccard, Auguste

UNITED STATES
Agassiz, Alexander
Agassiz, Louis
Bache, Alexander Dallas
Ballard, Robert Duane
Bass, George

Beebe, Charles William
Bjerknes, Jakob Aall
    Bonnevie
Bowditch, Nathaniel
Bowie, William
Brookes, William Keith
Bush, Katherine
Carpenter, William Benjamin
Carson, Rachel Louise
Cushman, Joseph
Dana, James Dwight
Dietz, Robert Sinclair
Feely, Richard Alan
Ferrel, William
Franklin, Benjamin
Heezen, Bruce Charles
Hess, Harry Hammond
Johnson, Douglas Wilson
Kofoid, Charles Atwood
Loeb, Jacques
Maury, Matthew Fontaine
McCammon, Helen Mary
Medwin, Herman
Menard, Henry William
Morse, Edward Sylvester
Munk, Walter Heinrich
Pourtalès, Louis-François de
Rathbun, Mary Jane
Revelle, Roger Randall
    Dougan
Scranton, Mary Isabelle
Smith, Deborah K.
Smith, Sidney Irving
Sverdrup, Harald Ulrik
Tennent, David Hilt
Tharp, Marie
Verrill, Addison Emery
Von Damm, Karen L.

**5TH CENTURY B.C.E.–4TH
CENTURY B.C.E.**
Aristotle

5TH C.E.–15TH C.E.
Bede

1500–1699
Cardano, Girolamo
Halley, Edmond
Maillet, Benoît de
Marsigli, Count Luigi
    Ferdinando
Moro, Antonio Lazzaro
Rondelet, Guillaume

1700–1799
Anning, Mary
Audouin, Jean-Victor
Bellingsgauzen, Faddei
Bernoulli, Daniel
Bory de St. Vincent, Jean-
    Baptiste-Geneviève-
    Marcellin
Bougainville, Louis-Antoine,
    comte de
Bowditch, Nathaniel
Buffon, Georges-Louis LeClerc,
    comte de
Chappe d'Auteroche, Jean-
    Baptiste

Cook, James
Cuvier, Georges, Baron
Dumont d'Urville, Jules-
    Sébastien-César
Duperrey, Louis-Isidore
Ehrenberg, Christian Gottfried
Forchhammer, Johann Georg
Forster, (Johann) Georg Adam
Forster, Johann Reinhold
Franklin, Benjamin
Geoffroy Saint-Hilaire,
    Étienne
Goethe, Johann Wolfgang von
Humboldt, Friedrich Wilhelm
    Karl Heinrich Alexander,
    baron von
Lacépède, Bernard Germain
    Étienne, de la Ville-sur-
    Illion, comte de
Lamarck, Jean-Baptiste-Pierre-
    Antoine de Monet,
    chevalier de
Lesueur, Charles-Alexandre
Lyell, Sir Charles
Poli, Giuseppe Saverio
Scoresby, William
Thompson, John Vaughan

1800–1809
Agassiz, Louis
Bache, Alexander Dallas

Brookes, William Keith
Darwin, Charles Robert
Desor, Pierre
Dujardin, Félix
Godwin-Austen, Robert
    Alfred Cloyne
Jeffreys, John Gwyn
Lenz, Heinrich Friedrich Emil
Lovén, Sven Ludwig
Maury, Matthew Fontaine
Milne-Edwards, Henri
Müller, Johannes Peter
Ross, Sir James Clark
Sabatier, Armand
Sars, Michael

1810–1819
Carpenter, William
    Benjamin
Dana, James Dwight
Ferrel, William
Forbes, Edward
Quatrefages de Bréau, Jean-
    Louis-Armand de
Wallich, George Charles

1820–1829
Leuckart, Karl Georg Friedrich
    Rudolf
Möbius, Karl August
Pourtalès, Louis-François de

1830–1839
Agassiz, Alexander
Dittmer, William
Hensen, Viktor
Morse, Edward Sylvester
Nordenskjöld, Nils Adolf Erik
Suess, Eduard
Thomson, Sir Charles Wyville
Verrill, Addison Emery

1840–1849
Albert I
Cleve, Per Teodor
Haeckel, Ernst Heinrich
     Philipp August
Lankester, Sir Edwin Ray
Makarov, Stepan Osipovich
Marion, Antoine-Fortuné
Murray, Sir John
Neumayr, Melchior
Renard, Alphonse
Smith, Sidney Irving

1850–1899
Bowie, William
Bush, Katherine
Delage, Yves
Lang, Arnold
Loeb, Jacques
Mohorovičić, Andrija
Murray, George Robert Milne
Penck, Albrecht
Richet, Charles Robert

1860–1869
Andrusov, Nikolai Ivanovich
Bjerknes, Vilhelm Frimann
     Køren
Charcot, Jean-Baptiste

Conklin, Edwin Grant
Goodrich, Edwin Stephen
Haldane, John Scott
Knipovitch, Nikolai
     Mikhailovitch
Kofoid, Charles Atwood
Nansen, Fridtjof
Portier, Paul
Rathbun, Mary Jane
Wegener, Alfred Lothar

1870–1879
Beebe, Charles William
Brookes, William Keith
Derugin, Konstantin
     Mikhailovich
Ekman, Vagn Walfrid
Helland-Hansen, Bjørn
Johnson, Douglas Wilson
Kylin, Johann Harald
Schmidt, Ernst Johannes
Tennent, David Hilt

1880–1889
Bailey, Sir Edward Battersby
Cushman, Joseph
Matsuyama (Matuyama),
     Motonori
Mesyatsev, Ivan Illarionivich
Pettersson, Hans
Piccard, Auguste
Sverdrup, Harald Ulrik
Vening Meinesz, Felix Andries
Zubov, Nikolay Nikolaevitch

1890–1899
Bjerknes, Jakob Aall Bonnevie
Holmes, Arthur
Wüst, Georg Adolf Otto

1900–1909
Ewing, William Maurice
Hess, Harry Hammond
Kuenen, Philip Henry
Revelle, Roger Randall
     Dougan
Shtokman, Vladimir
     Boresovich
Wilson, John Tuzo

1910–1919
Cousteau, Jacques
Hunt, Maurice Neville
Munk, Walter Heinrich
Schafer, Wilhelm

1920–1929
Heezen, Bruce Charles
Mann, Kenneth H.
Medwin, Herman
Menard, Henry William
Tharp, Marie

1930–1939
Bass, George
Earle, Sylvia Alice
Matthews, Drummond
     Hoyle
McCammon, Helen
     Mary
Vine, Frederick John

1940–1949
Ballard, Robert Duane

1950–1959
Scranton, Mary Isabelle
Smith, Deborah K.
Von Damm, Karen L.

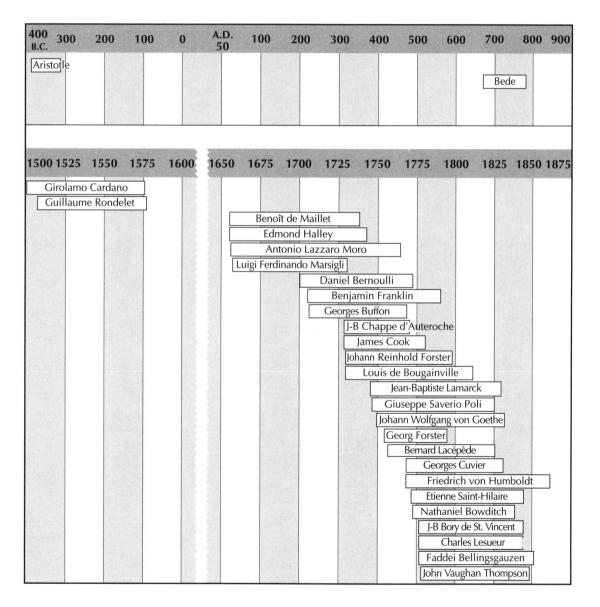

400 B.C.	300	200	100	0	A.D. 50	100	200	300	400	500	600	700	800	900

Aristotle

Bede

1500	1525	1550	1575	1600	1650	1675	1700	1725	1750	1775	1800	1825	1850	1875

Girolamo Cardano

Guillaume Rondelet

Benoît de Maillet

Edmond Halley

Antonio Lazzaro Moro

Luigi Ferdinando Marsigli

Daniel Bernoulli

Benjamin Franklin

Georges Buffon

J-B Chappe d'Auteroche

James Cook

Johann Reinhold Forster

Louis de Bougainville

Jean-Baptiste Lamarck

Giuseppe Saverio Poli

Johann Wolfgang von Goethe

Georg Forster

Bernard Lacépède

Georges Cuvier

Friedrich von Humboldt

Etienne Saint-Hilaire

Nathaniel Bowditch

J-B Bory de St. Vincent

Charles Lesueur

Faddei Bellingsgauzen

John Vaughan Thompson

1780	1800	1820	1840	1860	1880	1900	1920	1940	1960	1980	2000

Louis-Isidore Duperrey

William Scoresby

Jules Dumont d'Urville

Johann Georg Forchhammer

Christian Gottfried Ehrenberg

Jean-Victor Audouin

Sir Charles Lyell

Mary Anning

James Clark Ross

Henri Milne-Edwards

Johannes Peter Müller

Félix Dujardin

Heinrich Friedrich Emil Lenz

Michael Sars

Alexander Dallas Bache

Matthew Fontaine Maury

Louis Agassiz

Robert Alfred Cloyne Godwin-Austen

Charles Robert Darwin

John Gwyn Jeffreys

Sven Lovén

Jean-Louis-Armand de Quatrefages de Bréau

Pierre Desor

William Benjamin Carpenter

James Dwight Dana

Edward Forbes

George Charles Wallich

William Ferrel

Karl Georg Friedrich Rudolf Leuckart

Louis-François de Pourtalès

Karl August Möbius

Charles Wyville Thomson

Eduard Suess

Nils Adolf Erik Nordenskjöld

William Dittmer

Armand Sabatier

Alexander Agassiz

Viktor Hensen

Edward Sylvester Morse

Addison Emery Verrill

Per Teodor Cleve

Sir John Murray

Alphonse-François Renard

Ernst Heinrich Philipp August Haeckel

1780	1800	1820	1840	1860	1880	1900	1920	1940	1960	1980	2000

Sidney Irving Smith

Melchior Neumayr

Antoine-Fortuné Marion

Sir Edwin Ray Lankester

William Keith Brookes

Albert I (prince of Monaco)

Stepan Osipovich Makarov

Charles Robert Richet

Yves Delage

Arnold Lang

Katherine Bush

Edmund Beecher Wilson

Andrija Mohorovičić

George Robert Milne Murray

Albrecht Penck

Jacques Loeb

Alfred Lothar Wegener

Mary Jane Rathbun

Nikolai Ivanovich Andrusov

Fridtjof Nansen

Nikolai Mikhailovitch Knipovitch

Vilhelm Frimann Køren Bjerknes

Edwin Grant Conklin

Charles Atwood Kofoid

Paul Portier

Jean-Baptiste Charcot

Edwin Stephen Goodrich

John Scott Haldane

William Keith Bowie

David Hilt Tennent

Vagn Walfrid Ekman

Bjørn Helland-Hansen

Ernst Johannes Schmidt

Charles William Beebe

Konstantin Mikhailovich Derugin

Douglas Wilson Johnson

Johann Harald Kylin

Joseph Cushman

Sir Edward Battersby Bailey

Harald Ulrik Sverdrup

Motonori Matsuyama

Auguste Piccard

Ivan Illarionivich Mesyatsev

Nikolay Nikolaevitch Zubov

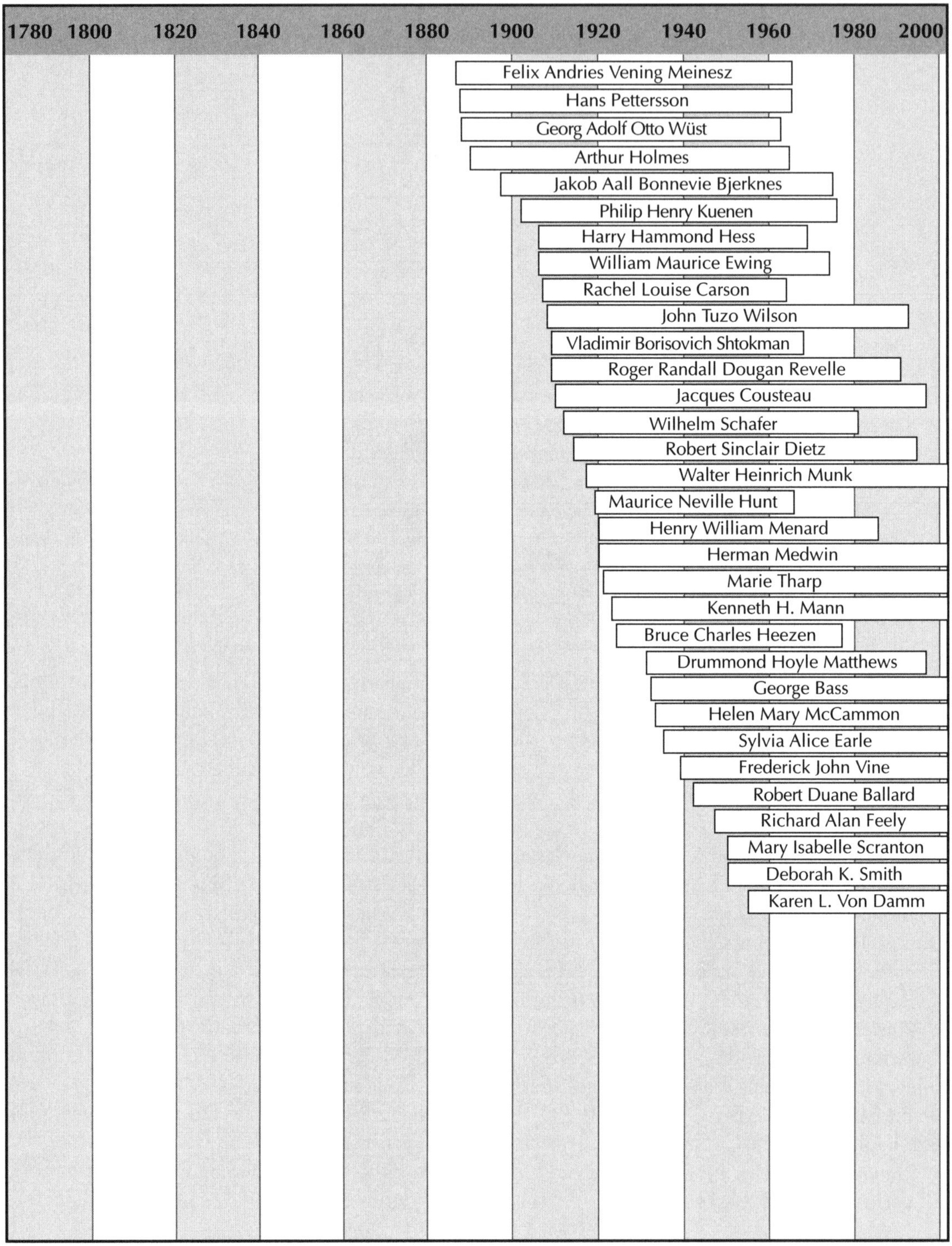

1780	1800	1820	1840	1860	1880	1900	1920	1940	1960	1980	2000

Felix Andries Vening Meinesz

Hans Pettersson

Georg Adolf Otto Wüst

Arthur Holmes

Jakob Aall Bonnevie Bjerknes

Philip Henry Kuenen

Harry Hammond Hess

William Maurice Ewing

Rachel Louise Carson

John Tuzo Wilson

Vladimir Borisovich Shtokman

Roger Randall Dougan Revelle

Jacques Cousteau

Wilhelm Schafer

Robert Sinclair Dietz

Walter Heinrich Munk

Maurice Neville Hunt

Henry William Menard

Herman Medwin

Marie Tharp

Kenneth H. Mann

Bruce Charles Heezen

Drummond Hoyle Matthews

George Bass

Helen Mary McCammon

Sylvia Alice Earle

Frederick John Vine

Robert Duane Ballard

Richard Alan Feely

Mary Isabelle Scranton

Deborah K. Smith

Karen L. Von Damm

Abbott, David. *The Biographical Dictionary of Scientists*. 3d ed. New York: Wiley, 1982.

——. *The Biographical Dictionary of Scientists: Biologists*. New York: P. Bedrick Books, 1983.

——. *The Biographical Dictionary of Scientists: Chemists*. P. Bedrick Books, 1984.

American Men and Women of Science, 1995–6: A Biographical Directory of Today's Leaders in Physical, Biological and Related Sciences. New York: R.R. Bowker, 1994.

Anderson, Paul. *Imagined Worlds: Stories of Scientific Discovery*. London: British Broadcasting Corp., 1985.

Anton, Ted. *Bold Science: Seven Scientists Who Are Changing Our World*. New York: W.H. Freeman, 2000.

Arago, François. *Biographies of Distinguished Scientific Men (1859)*. Freeport, N.Y.: Books for Libraries Press, 1972.

Asimov, Isaac. *Asimov's Biographical Encyclopedia of Science and Technology: The Lives and Achievements of 1510 Great Scientists from Ancient Times to the Present Chronologically Arranged*. New York: Doubleday, 1982.

Bailey, M. J. *American Women in Science*. Santa Barbara, Calif.: ABC-CLIO, 1994.

Betz, Paul, and Mark C. Carnes, eds. *American National Biography*. New York: Oxford University Press, 1999–2002.

Biographical Memoirs. Washington, D.C.: National Academy of Sciences, 1940– .

Bragg, Melvyn. *On Giants' Shoulders: Great Scientists and Their Discoveries: From Archimedes to DNA*. New York: Wiley, 1998

Bridges, Thomas, and Tiltman, Hubert H. *Master Minds of Modern Science*. New York: L. MacVeagh, 1931.

Concise Dictionary of Scientific Biography. New York: Scribner's, 2000.

"Cumulative Bibliography on the History of Oceanography." Available on-line. URL: *http://scilib.ucsd.edu/sio/indexes/cbho.html*. Downloaded January 31, 2003.

Current Biography. New York: H. W. Wilson, 1940– .

Daintith, John, et. al., eds. *Biographical Encyclopedia of Scientists*. Philadelphia: Institute of Physics Publishing, 1994.

Debus, Allen G. *World Who's Who in Science: A Biographical Dictionary of Notable Scientists From Antiquity to the Present*. Chicago: Marquis Who's Who, 1968.

Defries, Amelia. *Pioneers of Science*. London: Routledge, 1928.

Dictionary of American Biography. New York: Scribner's, 1928– .

Dictionary of Scientific Biography. New York: Scribner's, 1970– .

Downs, Robert B. *Landmarks in Science: Hippocrates to Carson.* Littleton, Colo.: Libraries Unlimited, 1982.

Elliot, Clark A. *Biographical Dictionary of American Science: The 17th Through the 19th Centuries.* Westport, Conn.: Greenwood Press, 1979.

50 Years of Ocean Discovery: National Science Foundation 1950–2000. Washington, D.C.: National Academy of Sciences, 2000.

Gascoigne, Robert M. *A Chronology of the History of Science, 1450–1900.* New York: Garland, 1987.

———. *A Historical Catalogue of Scientists and Scientific Books: From the Earliest Times to the Close of the 19th Century.* New York: Garland, 1984.

Greenstein, George. *Portraits of Discovery: Profiles in Scientific Genius.* New York: Wiley, 1998.

Hammond, Allen L. *A Passion to Know: Twenty Profiles in Science.* New York: Scribner's, 1984.

Jones, Bessie Zaban, ed. *The Golden Age of Science: Thirty Portraits of the Giants of 19th-Century Science by Their Contemporaries.* New York: Simon & Schuster, 1966.

Kass-Simon, Gabriele, and Patricia Farnes. *Women of Science: Righting the Record.* Bloomington: Indiana University Press, 1990.

Laidler, Keith J. *To Light a Candle: Chapters in the History of Science and Technology.* New York: Oxford University Press, 1998.

Late 17th Century Scientists. New York: Pergamon, 1969.

Meadows, Arthur J. *The Great Scientists.* New York: Oxford University Press, 1987.

McGraw-Hill Modern Scientists and Engineers. New York: McGraw-Hill, 1980.

Mid-19th Century Scientists. New York: Pergamon, 1969.

Millar, Ian, ed. *Chambers Concise Dictionary of Scientists.* Edinburgh: University of Cambridge Press, 1989.

Miller, David, ed. *Cambridge Dictionary of Scientists.* New York: Cambridge University Press, 1990.

Murray, Robert H. *Science and Scientists in the 19th Century.* UK: Gregg, 1971.

Nobel Foundation, "Nobel Laureates." Available on-line. URL: *http://www.noble.se.* Downloaded October 20, 2002.

Oceanography in the Next Decade: Building New Partnerships. Washington, D.C.: National Research Council, 1993.

Olby, Robert C. *Late 18th Century European Scientists.* New York: Pergamon, 1966.

Pais, Abraham. *The Genius of Science: A Portrait Gallery.* New York: Oxford University Press, 2000.

Pelletier, Paul A. *Prominent Scientists: An Index to Collective Biographies.* New York: Neal-Schuman, 1994.

Perutz, Max. *Is Science Necessary? Essays on Science and Scientists.* New York: Dutton, 1989.

Porter, Roy. *Man Masters Nature: 25 Centuries of Science.* New York: Braziller, 1987.

Scientific Genius and Creativity: Readings from Scientific American. New York: W.H. Freeman, 1982.

Scripps Institution of Oceanography Archives. Available on-line. URL: *http://scilib.ucsd.edu/sio/archives/.* Downloaded: October 20, 2002.

Shortland, Michael, and Richard R. Yeo. *Telling Lives in Science: Essays on Scientific Biography.* New York: Cambridge University Press, 1996.

U.S. Geological Survey. Available on-line. URL: *http://www.usgs.gov.* Downloaded: October 20, 2002.

Who's Who in Science and Engineering. New Providence, N.J.: Marquis, 1994.

Who's Who in Science in Europe. Essex, U.K.: Longman, 1994.

Who's Who of British Scientists. New York: St. Martin's, 1981.

Williams, Trevor I., *Collins Biographical Dictionary of Scientists*. 4th ed. Glasgow, Scotland: Harper Collins, 1994.

Woods Hole Oceanographic Institution. Available on-line. URL: *http://www.whoi.edu/*. Downloaded October 20, 2002.

INDEX

Note: Page numbers in **boldface** indicate main topics. Page numbers in *italic* refer to illustrations.